HARMONIC CHARTS

Understanding and using the principle of harmonics in astrological interpretation.

HARMONIC CHARTS

A New Dimension in Astrology

DAVID HAMBLIN M.A., D.F. Astrol.S.

THE AQUARIAN PRESS
Wellingborough, Northamptonshire

First published 1983

British Library Cataloguing in Publication Data

Hamblin, David
Harmonic Charts
1. Astrology
I. Title
133.5 BF1708.1

ISBN 0-85030-304-4

The Aquarian Press is part of the Thorsons Publishing Group

Printed and bound in Great Britain

CONTENTS

PART FOUR: OTHER APPLICATIONS

To my beloved Helen
and
to the memory of John Addey,
the founder of harmonic astrology

FOREWORD

During the weeks of his fatal illness in March 1982, John Addey struggled again and again to summon the energy and concentration to write a foreword to this important book. When, in his last hours, it became apparent that he was not going to be able to do so, one of John's final requests was to ask me to convey to David his great regret at not being able to write this foreword and to ask me to do it on his behalf. And it is indeed both a very great honour and pleasure to do so.

David Hamblin's work in the field of astrological interpretation is already well known to members of the Astrological Association and to readers of *The Astrological Journal*. In applying and developing the insights of John Addey on the all important role of Number in our understanding of the formative forces of human character and destiny, David reveals once again his great gift of interpretative insight. This enables him to get to the heart of a case history, and reveal the essential factors and motivations in terms of their corresponding harmonics.

In David, John recognized one of those very rare research workers who are able to marry sustained and patient study and gifted psychological insight with a deep practical sense, someone who is able to make theory come alive in the crucible of his or her own work and experience. John saw this book as a vitally important first link by which his theoretical ideas and empirical findings can become a standard part of the working tools of every serious astrologer, bringing the living energies of Ideal Numbers back into the heart of astrological practice.

If you have not yet begun to use harmonic charts in your interpretation then this book will open you to the delights and profound insights of what is indeed a truly 'new dimension in astrology'. To use harmonic charts is like moving from black-and-white photography to both colour and 3-D, at one and the same time. Your own familiar charts will suddenly take on entirely new perspectives. Fresh vistas will unfold, revealing within you, and those you thought you knew, unsuspected, yet suddenly obvious, dynamics and potentialities. To paraphrase the well-known beer advertisement, this technique truly 'reaches parts of the chart that no other methods reach'. But take care! Experience on our annual harmonic summer schools at Cambridge University, England, has shown that once students taste the power of these charts, they become addicted. Once you have discovered the harmonic dimensions of the horoscope anything less seems 'stale, flat, and unprofitable'.

To those who are suspicious of harmonics, imagining that they are an attempt to fit astrology into the world of 'science', in the narrow, contemporary, sense of that term, it should be said that their discoverer, John Addey, was by nature and training a mystic. He aspired not only to discover the truth about the place of astrology in the scheme of things, but also to express and apply these newly clarified first principles in such a way that astrology could become of even deeper practical value for the enlightenment and inspiration of mankind.

If John had survived to advise on the final manuscript of this book there is one point, I think which he would have wanted to suggest, a point he repeatedly made in his courses and lectures, This is the great potential value of relating the Ideal Numbers, which he saw as underlying harmonic theory, to the three great Ideals of classical philosophy – Goodness, Beauty and Truth – and the three main faculties of the human soul by which these are approached: the Will, Heart, and Mind. There is not space here to discuss these ideas in any detail, but the reflective student will find much food for thought in considering the first three prime numbers, 2, 3, and 5, in relation to the three Ideals and soul faculties in the order given. It was John's conviction that such an approach will eventually allow for the development of a true astro-psychology established on real First Principles. Meditation on such theoretical ideas may perhaps assist some to develop yet further the illuminating empirical material David has presented in these pages. You are about to embark upon a journey which will transform your astrology. Bon Voyage!

Charles Harvey

INTRODUCTION

John Addey – who died in 1982, when this book was already half-finished – is generally agreed to have been one of the greatest astrologers of the twentieth century; and it is also agreed that his greatest achievement was the introduction of harmonics into astrology. Thus, it is accepted that harmonics are one of the most important new developments – if not *the* most important new development – in astrology. Yet the proportion of astrologers who regularly use harmonics in their interpretative work is probably quite small. One reason for this is that there has been no book dealing with the detailed interpretation of harmonic charts. This book is an attempt to fill that gap.

The purpose of the book is to help the astrologer to understand harmonics and to introduce them into his or her practical work. I have assumed that the reader already has some understanding of basic traditional astrology, but I have *not* assumed that he or she already knows anything about harmonics. The basic concepts of harmonics, however, are not difficult: they are simply an extension of the principle of aspects, which has always been one of the cornerstones of astrology.

I believe that many astrologers have rejected harmonics, not only because of their supposed difficulty, but also because they are seen as belonging to the 'scientific' school of astrology. Those astrologers who see themselves as anti-scientific (more interested in intuition than in computers, or in Karma than in statistics) have tended to reject harmonics. I will try to show that this view is false. Even a glance at a harmonic chart can yield valuable intuitive insights which cannot be obtained from the traditional natal chart; if studied more closely the harmonic chart will yield a wealth of detailed information. Harmonics do not involve the rejection of traditional astrology, rather they provide a most valuable supplement to it at both the intuitive and the scientific level. In fact, I have come to regard the information to be obtained from harmonic charts as *indispensable* to true astrological understanding.

The book has been written as an aid to astrological interpretation and consultancy: it is *not,* therefore, a manual of research methods, and it will not provide direct help for the astrologer who wishes to carry out statistical research into the nature and meaning of harmonics. For this, one should read John Addey's *Harmonics in Astrology*[1] and also the section on harmonics in Geoffrey Dean's *Recent Advances in Natal Astrology.*[2]

It may be objected that this book is premature, since much research still needs to be done into the meaning of harmonics. While this is undoubtedly true, I do not accept that it is too soon to write about the interpretation of harmonic charts. If we had to wait for scientific proof, we would have to reject almost the whole of astrology except for the small amount which has been proved correct by the Gauquelins; and the conclusions reached in this book seem to me to be as well founded as most of the tenets of traditional astrology (on which, indeed, they are mostly based). On the other hand, it is also true that further research and further practice will lead to developments in the interpretation of harmonics which will render this book out-of-date. But this book is not meant to be the final answer: it is intended as one step along the road which will lead to the establishment of harmonics as a fully integrated part of astrological theory and practice.

The book is itself based on research, but it is research of a rather unusual kind. My basic method has been to study the harmonic charts of a large number of famous and not-so-famous people whole birth charts I have collected, magpie-like, ever since I first became interested in astrology more than ten years ago, and to look for interpretations of their harmonic charts which made sense in every case. Thus, for instance, if I found the same harmonic aspect in the charts of both Margaret Thatcher and Sigmund Freud, I would have to ask myself what these two people had in common, and in what way the aspect could be interpreted so as to make sense when applied to both of them. In some cases (where one lacks detailed knowledge about the lives and personalities of the people in question) one can only ensure that the interpretation is plausible; but in other cases the process can provide real insights which simultaneously illuminate both the personalities involved and the astrological theory.

The statements in this book are thus empirically based, but they are not scientifically based: that is to say, I have not 'proved' anything, and my interpretations are subjective and are dependent on the quality of my own insight and understanding. (Most of all, they are dependent on my articulateness, a quality in which I often feel sadly lacking: one can often 'see' intuitively the meaning of a particular configuration, but the problem is to find the *words* with which to express that meaning adequately.) I have not been able to present all the evidence on which the conclusions are based, since this would have meant filling the book with an enormous number of harmonic charts; but I hope that readers will make their own collections of harmonic charts, and will draw their own conclusions, which may often be different from mine (or at least more clearly expressed).

The method of interpreting harmonic charts which is presented in this book is a personal one, and is different in some respects from the method which was taught by John Addey and Charles Harvey in their annual Harmonics Courses at Cambridge. Thus (for reasons which I will explain in Chapters 19 and 21) I have placed no emphasis on the signs occupied by the planets in harmonic charts, or on aspects between harmonic planets and radical planets, or on 'harmonic directions'; but, on the other hand, I have placed a greater emphasis on the *patterns of aspects* between harmonic planets, which for me are the strongest clues for interpretation. Also my interpretation of some of the harmonic numbers (especially Five) is slightly different from John's.

But, despite these differences of emphasis, I must make it clear that my debt to John Addey, the founder of harmonic astrology, is of course enormous, and my admiration for him is equally great. John was the pioneer: the rest of us can only stumble along the road which he opened up for us.

I must also express my debt to Charles Harvey, whose Cambridge lectures on harmonics I found most valuable, and who has encouraged me throughout the writing of the book and has consented to write the Foreword. Others who have helped me include Dianne Binnington, Donald Stallybrass, and above all my wife Helen, who, in addition to her patience during the long period when the book was being written, has helped me with the diagrams and with the cover design.

An apology

I would like to end this Introduction by apologizing to women readers for the fact that (although there are three female case-studies in Part Three) the examples given in Chapters 4-7 are almost entirely male. Despite this, the proposals about the interpretation of harmonic charts are of course intended to apply to women just as much as to men. Throughout the book, the words *he, him, his* (except where they refer to specific individuals) should be taken to mean *he/she, him/her, his/her*; and the word *man* refers to the whole of mankind – women and men.

I would also like to apologize to 'Linda' and 'Susan', whose charts I have used in Chapters 15 and 16. It is of course usual to ask for clients' permission to use their charts for publication, even when anonymity is preserved; but in the case of Linda and Susan (who both consulted me more than seven years ago) I have no record of their addresses and have been unable to contact them. I hope that, if either Linda or Susan should read this book, they will forgive me my impertinence in using their charts (and also in using Linda's excellent 'autobiography'), and will obtain some pleasure from reading about themselves.

PART ONE: BASICS

Part One is concerned with basic issues: what harmonics are; which harmonic charts are most important; and how to calculate and present a harmonic chart.

CHAPTER 1
WHAT ARE HARMONICS?

The purpose of this book is to introduce the reader to an exciting new technique in astrology: the use of harmonic charts. For each birth chart, a series of harmonic charts can be worked out. Each of these charts shows the individual in a new light, and shows a different aspect of his perception of reality. To look at a series of these charts, alongside the birth chart itself, is like peeling away the different layers of the person's soul. Through the use of harmonic charts, it is possible to obtain a far fuller and more rounded picture of the individual's personality than can be obtained from the birth chart alone.

Thus, harmonic charts are an invaluable aid to astrological interpretation and consultancy. But harmonics are more than an aid. By studying harmonics, we obtain a new insight into the essential meaning of the birth chart itself, and into the real principles on which astrology is founded. And beyond this again, we begin to obtain an insight into the true meaning of Number, and thus into the principles of the universe itself.

I have been a devotee of harmonics ever since I first heard John Addey speak on the subject, two or three years before the publication of his book *Harmonics in Astrology*[1] in 1976. Many astrologers who were around in the mid-70s will remember the eagerness and impatience with which John's book was awaited, and the rapturous reception which it received when it appeared. Here indeed was a real breakthrough, a radical departure which, in Sheila Geddes's words, would ensure that 'astrology will never be the same again'.[2]

And yet, in a sense, harmonics are not a departure or a new beginning at all. This is not only because harmonics have always been used in Hindu astrology, but also because the principles underlying harmonics are those which have always been at the foundation of Western astrology. Unlike other new (or relatively new) astrological techniques, such as midpoints and composite charts, harmonics do not involve the introduction of any new basic principles. They simply involve the extension of existing principles to a wider field.

In order to understand this, we must go back to first principles and ask what harmonics essentially are. However, in this book I will not discourse at length on the underlying theory of harmonics, since John Addey has done this so ably in his book. (Readers who have not already read John Addey's book are urged to do so.) Here I will only attempt to 'de-mystify' harmonics by explaining what a harmonic chart is, and what kinds of insights can be obtained from its use.

Harmonics are essentially to do with *aspects*. They are no more nor less than a means of studying a wider range of aspects than those used in traditional astrology. In studying harmonics, then, we can start by looking at the traditional aspects and making sure that we understand their meaning, before going on to look at the new types of aspect which are introduced by harmonics.

In traditional astrology, the most important aspect is the conjunction. When two planets are conjunct in the birth chart, we regard them as being united or merged with each other. The relationship between them, in other words, is of the nature of *Oneness* or unity.

The next group of traditional aspects consists of those which are regarded as the 'hard' or 'difficult' aspects: the opposition, the square, and the semi-square.* In the opposition, planet B is half-way (½) round the circle from planet A; in the square, it is one-quarter (¼) of the way round; and in the semi-square it is one-eighth or three-eighths (⅜) of the way round. But 4 is 2 × 2, and 8 is 2 × 2 × 2. Therefore these are all aspects which are of the nature of *Twoness*. Astrologers (both Western and Eastern) have always believed that they undestand what Twoness means. Twoness is to do with tension, conflict, challenge, difficulty, effort. All these aspects, therefore, have this character, since they are all 'Twoness' aspects. They show that the relationship between planet A and planet B is of a Two nature, that it is a difficult, tense, challenging relationship.

Next we have the trine, which is regarded as a 'soft' or 'easy' aspect. In the trine, planet B is one-third (⅓) of the way round

* There is also the sesquiquadrate, which is three-eighths of the circle. Since 'sesquiquadrate' is a long and difficult word, and since there is in my opinion no interpretative difference between the semi-square and the sesquiquadrate, I will in this book use the term 'semi-square' to include sesquiquadrates as well as the traditional semi-squares.

the circle from planet A. The trine is therefore an aspect of pure *Threeness*. Once again, astrologers have proceeded on the assumption that they understand what Threeness means: Threeness is to do with harmony, pleasure, enjoyment, facility, 'free flow'. The relationship between planet A and planet B is of the nature of Threeness.

Finally we have a group of aspects which have an essentially 'mixed' nature: the sextile, the semi-sextile, and the quincunx. The sextile (one-sixth of the circle) is regarded as being *mainly* a 'soft' aspect like the trine, but it is recognized that since 6 is 2 × 3, an element of Twoness (tension, effort, etc) is added to the pure facility and enjoyment of the trine: with the sextile, the person is *striving towards* facility and enjoyment. (Or, to put it the other way round, he is enjoying the experience of conflict and difficulty.) The semi-sextile and the quincunx represent one-twelfth and five-twelfths respectively, † and, since 12 is 2 × 2 × 3, it is recognized that the element of conflict and effort is stronger in these aspects than in the sextile and that these aspects are therefore more 'difficult': they have more Twoness in them than the sextile, and therefore less Threeness. But they are still, like all the traditional aspects except the conjunction, built up from the numbers Two and Three.

More recently, some astrologers have started looking for quintiles (one-fifth of the circle) and bi-quintiles (two-fifths) and entering them on the natal chart. These are aspects which partake of the nature of *Fiveness*. But this introduces a new question: what is Fiveness, and how does it differ from Twoness and Threeness? And, if we are prepared to admit quintiles, why should we not also look for semi-quintiles (one-tenth of the circle), which combine the nature of Twoness and Fiveness? Why should we not look for aspects which combine Threeness and Fiveness? Why not look for Seven and other prime numbers? Why not (following Edith Wangemann[3]) look for further combinations of Two and Three, such as twenty-fourths and forty-eighths? Where will it all end? Is there any limit to the number of meaningful aspects which can be entered on the natal chart?

† Once again I am inclined to believe that, for interpretative purposes, the semi-sextile and the quincunx are exactly the same, since they are both 'Twelve' aspects. But here I recognize that I am on controversial ground, since many astrologers regard the quincunx as a stronger aspect, and also as more 'difficult', than the semi-sextile.

It is, of course, possible to enter on the birth chart far more aspects than those used in traditional astrology. But here we come up against a difficulty, which is the result of the limitations of the human brain. When a great deal of very diverse information is presented on a single sheet of paper, the brain has considerable difficulty in comprehending it as a whole and seeing the *patterns* within it. If the amount of information is too great to be presented on a single chart, it becomes worthwhile to draw a number of different charts, each concentrating on a particular type of information.

This, then, is the justification for drawing a number of harmonic charts for a single birth chart: each of the harmonic charts concentrates on a particular type of aspect. The practice of drawing harmonic charts can be compared to the practice of drawing a number of different geographical maps of the same area: a political map, a contour map, a land-use map, a weather map, and so on. If all this information was presented on a single map, the brain would not be able to comprehend it. But if it is presented on a number of different maps, then the maps can be used to supplement each other and to build up a total picture in the mind.

What, then, is a harmonic chart? In drawing a harmonic chart, we are dividing the circle of the zodiac into a number of equal segments, and treating each segment *as if* it was a whole circle. In drawing the fifth-harmonic (H5) chart, we are dividing the circle by 5; in drawing the seventh-harmonic (H7) chart, we are dividing it by 7; and so on.

As an example, we can take the fifth-harmonic (H5) chart. To set up an H5 chart, we start by dividing the circle of 360 degrees (starting at 0° Aries) into five equal segments. The first segment is from 0° Aries to 12° Gemini, the second from 12° Gemini to 24° Leo, the third from 24° Leo to 6° Scorpio, the fourth from 6° Scorpio to 18° Capricorn, and the fifth from 18° Capricorn back to 0° Aries. Each of the segments has an angular width of 72 degrees.

Having done this, we treat each of the segments *as if* it contained a whole zodiac of 360 degrees. This means that we multiply the angular distances within the segments by 5. A planet at 0° Aries is still at 0° Aries, but a planet at 1° Aries is now at 5° Aries, and a planet at 2° Aries is now at 10° Aries. (Similarly, a planet at 0°08′ Aries is now at 0°40′ Aries.) The same applies to the second segment (the one that starts at 12° Gemini) and to all the other segments. A planet at 12° Gemini is now at 0°

Aries, a planet at 13° Gemini is now at 5° Aries, a planet at 14° Gemini is now at 10° Aries, and so on.

Thus we can see that a planet whose radical position is 1° Aries will appear on the H5 chart as *conjunct* with a planet whose radical position is 13° Gemini. Both will appear on the H5 chart at 5° Aries. The aspect between these two planets in the radical chart (the birth chart itself) is a quintile (one-fifth of the circle), but in the H5 chart quintiles (and bi-quintiles) are converted into conjunctions. Similarly, semi-quintiles (one-tenth and three-tenths of the circle) are converted in the H5 chart into oppositions, and semi-semi quintiles (one-twentieth, or any larger number of twentieths) are converted into squares. Likewise, aspects representing fifteenths of the circle are converted into trines, and so on.

Thus, the H5 chart is a map of the *way in which the planets in the radical chart resonate to the nature of Fiveness*: or, more simply, it is *map of the Fiveness* in a person's astrological personality. Aspects with the nature of pure Fiveness are shown as conjunctions in the H5 chart; aspects showing a combination of Fiveness and Twoness are shown as oppositions, squares and semi-squares; and aspects representing a combination of Fiveness and Threeness are shown as trines. (Similarly, we can draw sextiles, semi-sextiles and quincunxes, which represent combinations of Fiveness, Threeness and Twoness.) Since we already understand something about the meaning of Twoness and Threeness, we can already say that the 'hard' aspects in the H5 chart will show the person making an *effort* towards Fiveness, whereas trines will show him having an *enjoyment of, or facility in,* Fiveness. But what is Fiveness, and what principles does it represent? We will try to answer this question in Chapter 5, when we look at the interpretative meaning of the H5 chart.

The same applies to any other harmonic chart. Thus, in drawing up the ninth-harmonic (H9) chart, we divide the zodiac into nine equal segments and treat each segment *as if* it contained a whole zodiac. The resulting chart will show the way in which the planets (and the other features of the chart) resonate to the nature of Nineness.

We can now see how the harmonic chart, although it is only a device, is extremely useful and valuable. It allows us to *separate off* the Fiveness in the chart from all the other features, so that we can study it on its own. It enables us to introduce a further level of complexity without being overwhelmed by it. If we can succeed in understanding the nature of Fiveness, we can understand from the H5 chart how a particular person will demonstrate Fiveness in his or her personality.

But we must never forget that the harmonic chart *is* only a device; it is only a way of presenting certain features that were really present in the radical chart itself. A planet may show itself in the H5 chart at 5° Aries, but *really* (in the radical chart) it is at 13° Gemini. It may show itself as being on the Ascendant, but *really* (in the radical chart) it is not on the Ascendant but quintile to it. Therefore, in interpreting a harmonic chart, we must constantly refer back to the radical chart. The planets that are shown on the harmonic chart are still the same planets that were shown on the radical chart, and their true position is still the same.

I believe that the radical chart is more important than any of the harmonic charts: but the harmonic charts can provide invaluable additional insights. The importance of using harmonic charts in consultancy is that if the principle of Fiveness is strong in the client's personality, we cannot discover this without drawing his H5 chart; and, if we fail to discover it, we may be omitting an essential part of his personality, without which we cannot fully understand him. Of course, the complexities of the human personality are so great that *full* understanding is beyond our reach. But the harmonic charts can provide some of the vital clues which we need in order to make sense of our perceptions and advise the client accordingly.

CHAPTER 2
WHICH HARMONICS?

A harmonic chart can be drawn for any number, and the number of harmonic charts that can be drawn for any one birth chart is therefore unlimited. But to draw an infinite number of harmonic charts would take an infinite amount of time, and we must therefore make a rational selection of those harmonic charts which are most likely to be useful in interpretation. For the practising astrologer, the selection will have to be ruthless: he will not want to calculate and draw even one harmonic chart unless he can reasonably expect to find it useful.

My recommendation is that the reader should concentrate on four particular harmonic charts: the fourth, fifth, seventh and ninth harmonics. (Readers who wish to experiment with other harmonics are, of course, welcome to do so; but these are the ones which I have found most useful.)

The argument for the fifth and seventh harmonics is simply that Five and Seven are the first two prime numbers beyond Two and Three, and Fiveness and Sevenness are therefore the first two new principles which are not revealed by the ordinary birth chart. By drawing the H5 and H7 charts, we can obtain an insight into the Fiveness and Sevenness in the individual's personality. In fact, for any readers who feel that two harmonic charts are quite enough to start with, I would recommend the fifth and seventh harmonics as being the most valuable.

The argument for the fourth- and ninth-harmonic charts is different. Four and Nine are not prime numbers but are the squares of Two and Three respectively. Thus, by drawing the H4 and H9 charts, we do not gain an insight into any new principles, but we do obtain a picture of pure Twoness and of pure Threeness, isolated from the other features of the chart.

There is some doubt about which is the best harmonic chart for showing the principle of pure Twoness. I have experimented with both the fourth- and the eighth-harmonic charts, and have come to the conclusion that the fourth-harmonic chart is more useful. The H4 chart also has the advantage that it is the one used in Reinhold Ebertin's midpoint system[1], so that those readers who wish to use midpoints as well as harmonics can obtain the midpoints directly from the H4 chart.

The H9 chart not only describes pure Threeness, but takes us to a level of Threeness which we would not otherwise penetrate, because the novile (one-ninth of the circle) is not an aspect which is ordinarily used in astrology. The H9 chart can therefore yield insights which are just as new as those obtained from the H5 and H7 charts.

In Chapters 4-7 I will discuss the interpretative meaning of the H4, H5, H7 and H9 charts. In Chapter 20, I will say something about the meaning of some other harmonic charts.

CHAPTER 3

CALCULATION AND PRESENTATION

In this chapter, I will describe the procedure for calculating and drawing harmonic charts, and introduce the principle of *harmonic orbs,* which is central to the argument of the book.

Calculation

There are four ways of calculating the positions of the planets and Angles in harmonic charts:

1 Use a computer. There are now a number of computer agencies that will calculate harmonic positions; or you can buy your own computer, with a programme for calculation of harmonics. (At the time of writing, the situation on computers is extremely fluid.)

2 Obtain a copy of the *Harmonic Chart Tables* by James S. Williamsen.[1] The position of a planet on any harmonic up to the 180th can be easily read off from these tables. The method is very simple, and is probably the easiest for the beginner to learn and apply.

3 Calculate the harmonic positions by hand. The best method is Method 3 as described by John Addey in Chapter 12 of *Harmonics in Astrology.*[2] It is tedious and time-consuming (although it becomes quicker with practice), and the risk of error is always present. However, for those readers who refuse to rely on mechanical or published aids, the method is set out in Appendix I.

4 Calculate the harmonic positions with a pocket calculator. This is the method I use, as I find it even quicker and more reliable than method 2 above. However it is important to have a calculator which has a key for converting from decimals into degrees (or hours) minutes and seconds, and back again into decimals. On my calculator the key for transferring to degrees, minutes and seconds is marked HMS; to perform the reverse

operation, I press two keys marked INV and HMS. Armed with such a calculator, the procedure is as follows:

1 Convert the planet's zodiacal longitude into *absolute* longitude, starting from 0° Aries (e.g. Taurus 15°53′ becomes 45°53′). A table for conversion into absolute longitude is given in Appendix II, and with practice you will soon memorize it.
2 Enter this figure (e.g. 45.53) into the calculator.
3 Press keys as follows:

INV HMS × ? = HMS

(? is the number of the harmonic, e.g. 5 for the fifth harmonic.)
4 Convert back from absolute longitude into zodiacal longitude.

Thus, the fifth-harmonic position of a planet at Taurus 15°53′ is calculated as follows:

Taurus 15°53′ = 45°53′
45.53 INV HMS × 5 = HMS
which gives the answer 229.25.
229°25′ = Scorpio 19°25′

But if the resulting figure is greater than 360, it is necessary to subtract 360, and to go on subtracting 360 until one has a figure which is less than 360. Thus the keys to be pressed are as follows:

INV HMS × ? − 360 = = = HMS

Thus, the seventh-harmonic position of a planet at Sagittarius 26°58′ is calculated as follows:

Sagittarius 26°58′ = 266°58′
266.58 INV HMS × 7 − 360 = = = = = HMS
which gives the answer 68.46.
68°46′ = Gemini 8°46′

This procedure may seem complicated but is in fact very simple, and with practice can be carried out in less than fifteen seconds for any one planet. However, with a *programmable* calculator, it is possible to reduce still further the time taken to calculate harmonic positions, by devising programmes that will do the work for you. All that is then necessary is to enter the planet's absolute longitude and set the programme in motion.

Presentation

When the harmonic positions of the planets and the Angles have been calculated, they can be entered on an ordinary chart form, which must of course be clearly marked not only with the person's name but also with the number of the harmonic. This method is shown in all the charts which are presented later in this book.

Beginners may find it surprising that, in a harmonic chart, the Midheaven (M.C.) may fall anywhere in the chart in relation to the Ascendant. It may just as well be below the 'horizon' as above it. I prefer always to draw the Ascendant on the left of the chart, and to let the M.C. fall where it may. Some readers may however prefer to draw the M.C. at the top of the chart, and let the Ascendant fall where it may.

In an *even-numbered* harmonic chart (such as the H4 chart), the Descendant is not opposite but conjunct the Ascendant, and the I.C. is not opposite but conjunct the M.C. (Similarly, if the Nodes were included, the South Node would be not opposite but conjunct the North Node.) But this does not apply in an *odd-numbered* harmonic chart (such as the H5, H7 and H9 charts).

My view is that the mundane houses should not be entered on the harmonic chart, since they would be meaningless: the only meaningful houses are the ones in the radical chart. Certainly it would be impossible to use any of the Quadrant house systems (such as Placidus or Koch) in a harmonic chart; the only possible system would be the Equal house system.

Aspects: harmonic orbs

When the planets have been entered on the chart, the aspects between them can be calculated. But here it is necessary to introduce the principle of *harmonic orbs*. This is the principle that (in John Addey's words) '*the orb must diminish in direct proportion to the number of the harmonic,* that is the number by which we have divided the circle to get our aspect.[3] Thus, if the orb allowed for the conjunction is 12 degrees, the orb allowed for the opposition must be 6° (12 divided by 2), because the opposition is the aspect of the second harmonic (i.e. it results from dividing the circle by 2); and the orb allowed for the trine must be 4 degrees (12 divided by 3), because the trine is the aspect of the third harmonic (i.e. it results from dividing the circle by 3).

It is absolutely essential that harmonic orbs are used when we are working with harmonic charts. In fact, I strongly recommend that harmonic orbs should be used in the radical chart as well as in the harmonic charts. The reason for this is easily demonstrated. If we start with an orb of 12 degrees for the conjunction, then the orb allowed for the square is only 3 degrees (12 divided by 4). But when we come to draw the H4 chart (in which the squares of the radical chart become conjunctions), all the distances between the planets are multiplied by 4. Therefore the orb allowed for conjunctions in the H4 chart is 3 degrees × 4, which is once again 12 degrees. Thus harmonic orbs are necessary in order to ensure that we can use the same orbs in the radical chart and in all the harmonic charts. If we were to allow an orb of (say) 8 degrees for squares in the radical chart, then this would become an orb of 32 degrees for conjunctions in the H4 chart, which would plainly be too wide.

In the end, however, it comes down to what one feels to be 'right'. Astrologers who are accustomed to allowing the same orb for all the major aspects (conjunctions, oppositions, squares and trines) may well find it difficult to adjust to harmonic orbs. But I believe that, having tried them, they will come to feel that harmonic orbs give a truer picture of the relationship between planets than the traditional system, and they will want to use them in the radical chart as well as in the harmonic charts.

The precise orbs allowed depend, of course, on the orb which is decided on for the conjunction. Having experimented with a number of systems, I have come to believe that the orb allowed for the conjunction should be 12 degrees – both because 12 is a number which is easily divided by other numbers, and also because this width of orb appears to be 'about right' in terms of the numbers of aspects which it includes and the number which it excludes. (But, in the matter of orbs, perfection is impossible to obtain, since it is clear that an aspect becomes gradually weaker as it moves away from exactitude, and does not suddenly cease to function at any one point.)

With an orb for the conjunction of 12 degrees, the orbs allowed for the particular aspects are as follows:

Conjunction	12°
Opposition	6°
Trine	4°
Square	3°
Sextile	2°
Semi-square	1°30′
Semi-sextile } Quincunx }	1°

(These figures are taken from Addey's book[4], where Addey also gives the orbs that result from a conjunction-orb of 15 degrees. In this book, the figures for the conjunction-orb of 12 degrees are reproduced in Appendix III for ease of reference.)

Only these (traditional) aspects should be calculated, both in the radical chart and in each of the harmonic charts. (There is no need, for instance, to calculate quintiles, since quintiles in the radical chart will be visible as conjunctions in the H5 chart.)

I would now like to introduce a further complication, which readers may ignore if they wish, but which in my view is a great aid to interpretation. It is clear that aspects are much more powerful when they are exact, or very close to exactitude. It is therefore helpful to note which aspects in a chart are especially 'close'. But, if the orb for aspects varies with the harmonic number of the aspect, then the orb for 'close' aspects should vary according to the same principle.

Therefore an orb of 2 degrees should be allowed for 'close' conjunctions, and this should be reduced for the other aspects as follows:

Conjunction	2°
Opposition	1°
Trine	40′
Square	30′
Sextile	20′
Semi-square	15′
Semi-sextile } Quincunx }	10′

(These figures are again reproduced in Appendix III for ease of reference.)

Presentation of aspects

Aspects, once calculated, can be listed in the usual way at the bottom of the chart form. ('Close' aspects can be underlined, or marked with a C for 'close' or an E for 'exact'.) They should then be entered on the chart itself, by means of *aspect lines* connecting the planets.

Two colours should be used for aspect lines: colour A and colour B. (The question of which colours will be discussed at the end of this chapter.) In the charts in this book, because of the difficulties of colour reproduction, colour A will be indicated by thick lines and colour B by thin lines. Colour A indicates the principles of Twoness, and colour B indicates the principle of Threeness.

Oppositions and squares should be indicated by a continuous line in colour A.

Trines and sextiles should be indicated by a continuous line in colour B.

Semi-squares should be indicated by a *broken* line (i.e. a series of dashes) in colour A.

Semi-sextiles and quincunxes should be indicated by a *broken* line consisting of alternating dashes of colours A and B.

'Close' aspects should always be indicated by a *double* line (of the type appropriate to the nature of the aspect: e.g. a close semi-square would be indicated by a double broken line in colour A).

Planets in **conjunction** should be bracketed together by means of a short curving line (as drawn in the example charts later in the book). These 'conjunction brackets' should be shown in a third colour to represent Oneness.

This is the system which I have worked out by a long process of experimentation, and which I have found to be the best one for revealing the *patterns of aspects* in a harmonic chart (and also in the radical chart). The recognition of these patterns is of prime importance in the interpretation of harmonic charts, and will be discussed in Chapter 8.

The harmonic chart is now fully set up, and ready for interpretation. Two matters remain to be considered in this chapter: aspects to the Angles, and the use of colours in drawing harmonic charts.

Aspects to the Angles

If we were absolutely sure of the accuracy of the birth times we are using, aspects to the Angles would probably be very important in harmonic charts. But the Angles in harmonic charts must be treated with great caution because of the inaccuracy of most birth times (since, in the harmonic chart, the extent of the inaccuracy is multiplied by the number of the harmonic). Because of this I do not usually take note of aspects (squares, trines etc.) to the Angles.

However, conjunctions to the Angles may be worth noting, because (according to the principle of harmonic orbs) they are valid over a wider orb than the other aspects, and so are less affected by inaccuracies in the birth data. I therefore note (by visual inspection of the chart, rather than by listing) which planets are conjunct to any of the Angles (Ascendant, Descendant, M.C., I.C.) within an orb of approximately 12 degrees.

The use of colours in harmonic charts

Harmonic charts can be made more attractive, and also easier to read and interpret, if they are presented in colour, with different colours used to represent the different harmonic numbers. Readers who are happy with black and white may of course ignore this section, but for those readers who are interested, I will describe the colour system which I use.

I have already suggested that there should be two colours (colour A and colour B) to represent Twoness and Threeness, and a third colour to represent Oneness. But in the H5 and H7 charts we are also dealing with two other prime numbers, Five and Seven, so we need two more colours to represent Fiveness and Sevenness.

After much experimentation I have found that (for me at least) the most appropriate colours are as follows:

Oneness	black
Twoness	red
Threeness	golden-yellow
Fiveness	blue
Sevenness	violet

(For Fiveness, a light blue should be used, which is clearly distin guishable from violet. Threeness should be represented by a deep yellow, tending towards orange, but clearly distinguishable from red; pale yellow would not be visible against white paper.)

I cannot justify these colours except by saying that they seem to me to be aesthetically right, and to convey the qualities of the different harmonic numbers. The absence of green from the list may seem surprising, but green seems to be of its nature a static colour, inappropriate for conveying a dynamic principle.

All the basic features of the chart (the name at the top of the chart, the glyphs for the signs and the planets, the figures for the planetary positions, and the listings of aspects) should be drawn in the appropriate harmonic colour. For the H4 chart the appropriate colour is red, since 4 is 2 × 2, and for the H9 chart the appropriate colour is golden-yellow. Thus the list of the main colours used in the harmonic charts is as follows:

Radical chart	black
H4 chart	red
H5 chart	blue
H7 chart	violet
H9 chart	golden-yellow

However, I have found it useful to draw the four Angles of the chart, and also the glyphs and figures for the Sun and Moon, in a contrasting colour. The most appropriate contrasting colours seem to be as follows:

Radical chart	red
H4 chart	black
H5 chart	red
H7 chart	red
H9 chart	black

The conjunction-brackets should be drawn in the main harmonic colour (e.g. violet in the H7 chart). Other aspect lines (oppositions, trines, etc.) should be drawn in the same colours (colours A and B) in all harmonic charts, since they always represent Twoness and Threeness. If we follow this scheme, colour A should be red, and colour B should be golden-yellow.

If we use colours in this way, the different types of harmonic chart can be instantly distinguished from each other. An H7 chart, for example, *looks* like an H7 chart because of the predominance of violet, and it cannot be mistaken for any other type of chart. Its appearance conveys something of the quality of its contents, and this is an invaluable aid to interpretation.

PART TWO: INTERPRETATION

Part Two deals with the interpretation of harmonic charts. In Chapters 4-7 we will discuss the basic meaning of the fourth-, fifth-, seventh- and ninth-harmonic charts, giving examples of people who are unusually strong in each of these harmonics, and ending in each case with a suggested interpretation for each of the aspects (conjunction, opposition, etc) in the harmonic chart. In Chapter 8 we will make some general points about the way in which harmonic charts should be analysed and interpreted. In Chapter 9 we will discuss the meaning of each of the planets in harmonic charts. In Chapter 10 we will propose interpretations for aspects between each pair of planets in each of the harmonics; and in Chapter 11 we will propose interpretations for aspects between planets and the Angles.

CHAPTER 4
THE FOURTH-HARMONIC CHART

The fourth-harmonic (H4) chart is a map of a person's Twoness. We already understand a great deal about a person's Twoness from studying the oppositions, squares and semi-squares in his radical chart; but the H4 chart allows us to probe it more deeply.

Much of the information in the H4 chart was already present in the radical chart, but it is presented in a different way. Thus, all the conjunctions in the H4 chart were present in the radical chart as conjunctions, oppositions or squares; all the oppositions were present as semi-squares; and all the trines were present as trines, sextiles, semi-sextiles or quincunxes. But the squares, semi-squares, semi-sextiles and quincunxes in the H4 chart show new information that we could not glean from the radical chart.

In order to understand and interpret the H4 chart, we have to think about the nature of Twoness. Basically, Twoness is about awareness of a difference, or a need, or a lack. When two planets are opposite to each other (that is, in a 'Two' relationship to each other) in the radical chart, the person is aware that these two planets are pulling in different directions and that he *needs* to bring them together in order to relate successfully to the outside world. The integration of these two planets is something which he feels that he *lacks,* but which he must work towards.

If a third planet is square to the other two (forming a T-square), the awareness of lack will be even stronger, but at the same time the need to overcome it will be stronger also. The person will feel that he *lacks* the qualities of the third planet, and that it is this lack which prevents him from bringing the first two planets together. He will strive hard to develop the qualities of the third planet in order to integrate all three planetary forces in his behaviour.

In the H4 chart, all the planets in the radical T-square will be shown in a conjunction, which can be regarded as a *cluster of effort.* The conjunction of planets in the H4 chart shows, not that these planets are integrated in the personality, but that the person is striving to integrate them through his behaviour.

But the H4 chart may also contain oppositions (which are semi-squares in the radical chart): and these show planets whose inter-relationship has a quality of still profounder Twoness. For example, if Mercury and Mars are conjunct in H4, they may both be opposite to Saturn: and so the person may feel that it is his lack of Saturn-qualities which prevents him from integrating Mercury and Mars. But in fact, at this level of Twoness, the person often seems to fail to recognize this as a problem within himself: he projects it on to the world, and thinks (for instance) that the world is telling him that he cannot integrate the qualities of Mercury/Mars and Saturn. His response to this is to try to prove them wrong by his visible actions and achievements. Hence, oppositions in H4 seem to relate particularly to *manifestation.* If planets are conjunct in H4, they show a struggle within the personality; but if they are opposite, they show a struggle to manifest the combined qualities of the planets in visible action, perhaps in response to a challenge which the person feels has been thrown down to him by the world.

Beyond this, the H4 chart may contain squares (which are not visible in the radical chart). These represent the next level of Twoness: the person feels that a block has been placed in the way of manifestation, and that he must struggle to remove or overcome it. Here the tendency to projection is still greater: the person often seems to feel that there are particular people in the outside world who are trying to prevent him from manifesting a part of his personality, and that he must struggle to overcome these 'enemies' in order to achieve his ends. He identifies himself with one of the planets in the square, and attributes to his 'enemies' the qualities of the other planet. If there is a Grand Cross, with two planets square to two other planets, this tendency will be still more evident. (Henri Rousseau, whose case we will discuss in Chapter 13, is an excellent example of this.) But in fact the planet which he is projecting is really part of his own make-up, so that he may often (perhaps unconsciously) *provoke* his 'enemies' into displaying the qualities of the projected planet so that he may attack them. The conflict between planets that are square in H4 can only be resolved by *victory* over the supposed enemy. But it is also possible for the person to use these squares as an excuse for inactivity: since victory cannot be achieved, he has to lie low and adopt a defensive attitude in the face of the 'enemy's' hostility.

Beyond this, there may be semi-squares in the H4 chart.

Following the theory, this should mean that the person feels that a block has been placed in the way of victory: he is prevented by a particular planet from fighting his enemies as he feels he must. In practice, it seems likely that semi-squares in H4 are not very significant if they are isolated from other aspects; but if they are integrated into a pattern of oppositions and squares, then they may be highly significant as representing a further level of obstacle that the person feels he must overcome.

Thus, the more conjunctions, oppositions, squares and semi-squares there are in the H4 chart, the more the person will spend his life battling against the world, and struggling to prove that he can in fact demonstrate qualities which he feels that the world wants to deny him. The type of success which he wants is very much success in the world: he wants to beat the world on its own terms. The archetypal H4 person is Demosthenes, who became a brilliant orator in order to overcome his stammer: he became the best at the very thing at which he was worst. And yet it would be wrong to assume that H4 strivings always lead to brilliant achievements. Striving can lead to great success or to great failure: the common factor is simply the striving.

Examples of strong H4 charts

As an example of a successful person with a strong H4 chart, we can take **Eddy Merckx**, four-times winner of the Tour de France cycling race, whom Gauquelin[1] describes as 'the best example of the iron-willed champion'. Merckx has two separate patterns, both very strong, in his H4 chart (p.38). Firstly, he has Sun conjunct Moon (showing that he is essentially an H4 person, striving towards self-integration); but both Sun and Moon are opposite to Venus and Pluto, showing that he feels he lacks attractiveness (Venus) and ruthlessness (Pluto) and feels he must overcome these lacks and manifest these qualities to the world in order to achieve his Sun-Moon integration. This pattern alone would tend to drive him into the kind of self-punishing activity which would win him the adulation of the world. But also, he has Mars closely opposite Jupiter, with both square to Saturn (the Mars-Saturn square is exact, and so is a particularly compelling force), showing that he also feels a strong need to manifest the Mars-Jupiter quality of unbridled physical energy, and that in order to achieve this he has to overcome the obstacle of Saturn. Squares in H4, as we have said, tend to be externalized, and Merckx may well have projected Saturn on to the organizers

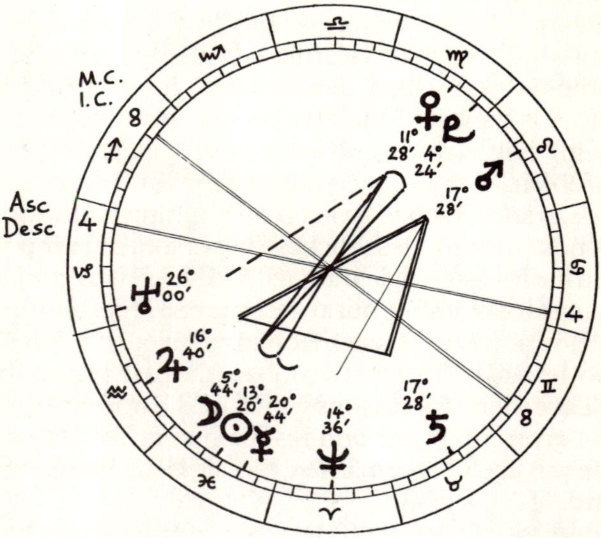

Eddy Merckx: H4 chart

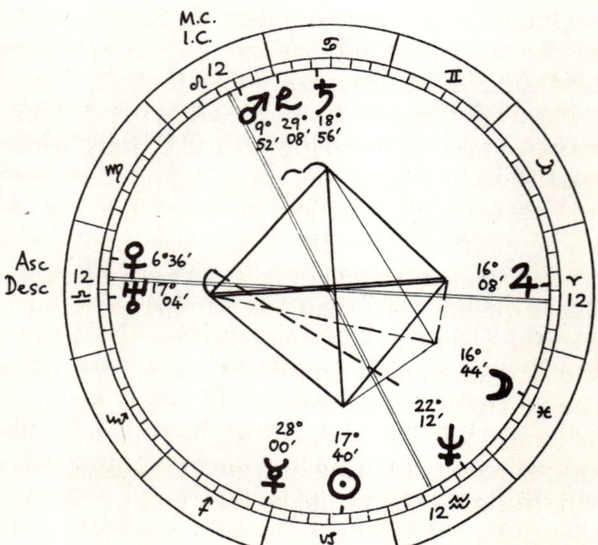

'A Tramp': H4 chart

of the race, the bureaucrats and hard-faced officials who tried to prevent him from achieving his objective. But it is clear that he needed this external Saturnian challenge in order to demonstrate his 'iron-willed' qualities and beat the world on its own terms. It is typical of such H4 people that they are never satisifed with their victories, which do not in fact resolve their internal problems: hence it was not enough for Merckx to win the Tour de France, he had to keep on winning and winning.

We can now turn to a less successful person, a **tramp** who once asked me for astrological advice. He had no fixed address and no changes of clothing, but at the time that I met him he was living in a Salvation Army hostel. He had never been able to hold down a job: he had had one or two jobs in the past, but had been forced to leave them, for reasons which he did not explain. But he had not given up hope, and his main concern was about whether he was likely to get married, though he had no particular girl in mind.

The tramp's H4 chart (p.38) shows a pattern which would make life very difficult. Firstly, there is the conjunction of Mars, Pluto and Saturn, which is the result of a T-square in the radical chart (Mars opposite Pluto, square to Saturn, which is conjunct the M.C.). But in the H4 chart, Saturn is at one of the corners of a complete Grand Cross. Saturn opposite the Sun shows that he can only resolve the Mars-Pluto-Saturn conflict by manifesting to the world the Sun-Saturn quality of self-control and self-limitation. But at right angles to the Sun-Saturn opposition is the very close opposition between Jupiter and Uranus, which is a further obstacle to be overcome. He will tend to externalize Jupiter and Uranus and to see them personified in the people around him: people who have all the bright, glowing, expansive, exciting qualities which he feels that he lacks (for how can he manifest these qualities and manifest Sun-Saturn at the same time?). Nevertheless, in his Saturnian way, he tries to manifest Jupiter-Uranus by adopting a completely independent life-style. He could be said to have been successful in his own way; but he is still unhappy, and it might be possible to advise him on a line of work which could bring him greater satisfaction. (For people with strong H4 charts, life will always be a battle, but a great deal depends on the *arena* in which the battle is fought out.)

I suspect in fact that it may be impossible to tell from the H4 chart whether the battle will result in success or failure. Often it may veer between the two. **Winston Churchill**, for example, had a strong but disjointed H4 chart (p.40), with four separate

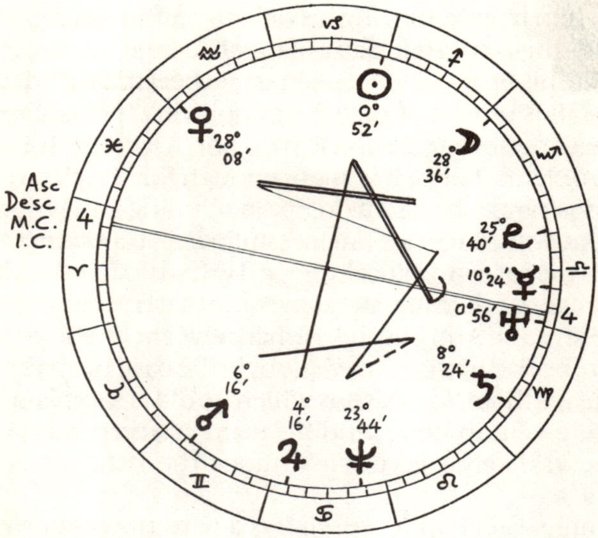

Winston Churchill: H4 chart

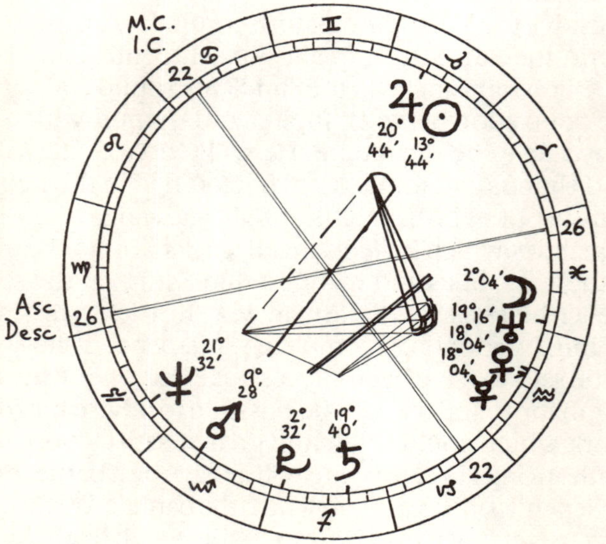

Franz Kafka: H4 chart

squares (two of them very close): Sun square Uranus, Moon square Venus, Mars square Saturn, and Neptune square Pluto. His life is counted as a great success: but if the Second World War had not occurred, he would have been written off as a failure.

Sometimes the person may be successful in one sense and unsuccessful in another. **Franz Kafka** was successful as a writer but unsuccessful in resolving his internal dilemmas. According to a critic quoted by Gauquelin,[2] Kafka 'wanted to be perfect and almost aimed at being a saint'. Gauquelin says that Kafka 'depicted in minute detail a daily life which he could hardly stand living. His very lucid, rational picture turns that life into the vision of a nightmare.'

For the sources of Kafka's creative imagination we must turn to harmonics other than the fourth; but the H4 chart (p.40) can tell us something about his problems in living. Anthony Storr says that

> Kafka, throughout his life, continued to attribute to his father almost magical powers and longed for an acknowledgment and recognition from him which he never felt able to obtain . . . . The feeling of the child's powerlessness compared with the parent persisted and impaired his whole adaptation.[3]

Hence it seems likely that Kafka viewed his father as the 'enemy' onto whom he projected the squares in his H4 chart. The chart shows that he has two different ways of viewing the 'enemy'. With his rational, feeling and inventive mind (Mercury/Venus/ Uranus square Jupiter) he sees his father as having the Jupiterian qualities of growth, power, brilliance, super-humanity: these are qualities which Kafka cannot obtain because the 'enemy' is keeping them from him. But Jupiter is conjunct the Sun, so until he has found Jupiter he cannot find himself. Hence the Sun becomes the father rather than the self, and Kafka cannot find his own identity separately from that of his father. On the other hand (Moon closely square Pluto), Kafka instinctively sees the 'enemy' as having qualities of ruthless destructiveness. The 'enemy' is not merely grand and powerful: he is also 'out to get him'.

I should add that Storr sees Kafka as a prime example of a schizoid personality. But I think it would be wrong to equate strength in H4 with schizophrenia: in my limited experience, schizophrenia is more usually associated with the absence of strong aspects.

This last case illustrates how the H4 struggles in which an adult engages are often the continuation and development of patterns of behaviour that were established during childhood. In most types of therapy it is recognized that we must go back to childhood in order to understand the root causes of an adult's present problems and neuroses, and the H4 chart can be an invaluable aid in this process. Of all the harmonics that we shall be considering, the fourth harmonic works closest to the surface. An individual can hardly help manifesting the characteristics of his H4 chart from his earliest years, whereas the other harmonics (especially the ninth) may remain latent.

Trines and sextiles in H4

So far, we have concentrated on the 'hard' aspects in the H4 chart: conjunctions, oppositions, squares and semi-squares. But the H4 chart can also contain 'soft' aspects: trines, sextiles, semi-sextiles and quincunxes. The H4 chart is a map of the person's Twoness, but the 'soft' aspects are aspects of Threeness, and so in the H4 chart they show a combination of Twoness and Threeness. If Twoness is effort and Threeness is pleasure, then these aspects show effort towards pleasure, or pleasure in effort: striving towards enjoyment, or enjoyment in striving. It is likely, then, that they will show a rather tense and uneasy kind of pleasure, as though the person was desperately trying to have a good time: and this interpretation is, in fact, often borne out.

For an example, we can take **Franchet d'Esperey**, a French Marshal in the First World War, known to the British as 'Desperate Frankey'. Gauquelin describes him as follows:

> His impetuousness in taking over direct command was typical: he drove cars at maniacal speed and shot windows when there was nothing better to shoot. He was known to terrorize his troops and his officers, and the experience of war had made him harsher, but people fought well under him, perhaps because he frightened them even more than the enemy did.[4]

D'Esperey's H4 chart (p.43) is extremely strong and contains a number of 'hard' aspects (Moon and Venus opposite Mars and Jupiter; Sun square Mercury). But even more striking is the group of trines and sextiles linking five planets: Sun, Venus, Mars, Neptune and Pluto. (The Sun-Pluto trine is virtually exact.) The Sun-Mars-Pluto Grand Trine shows a tense striving towards

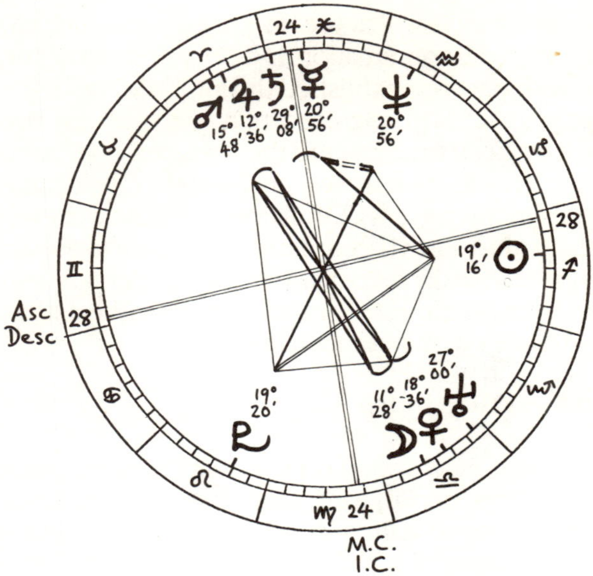

Franchet d'Esperey: H4 chart

enjoyment of ruthless activity, and this could well result in the maniacal behaviour which Gauquelin describes. The sextiles involving Venus and Neptune are more difficult to interpret: I am inclined to think that Venus-Neptune represents the idealized admiration which he felt was owed to him by his troops, and which he was striving to enjoy.

A weaker H4 chart

Finally we can look at an H4 chart which contains a less strong pattern of aspects. **Lucien Van Impe** is another cyclist who won the Tour de France, but who is described by Gauquelin as entirely lacking the 'will to win'.[5] He won through luck and through the skill of his trainer. His trainer said of him: 'He's never taken the initiative in anything and has never had any responsibilities. He's happy-go-lucky, a child happy with whatever turns up. No worries.' This would seem to be a personality which lacks the quality of H4 striving.

Van Impe's H4 chart is, indeed, not very strong (see p. 44). (The close Mercury-Mars conjunction was a conjunction in the radical chart also, and so does not have an H4 quality. And aspects which involve only the outer planets – Jupiter, Saturn, Uranus, Neptune

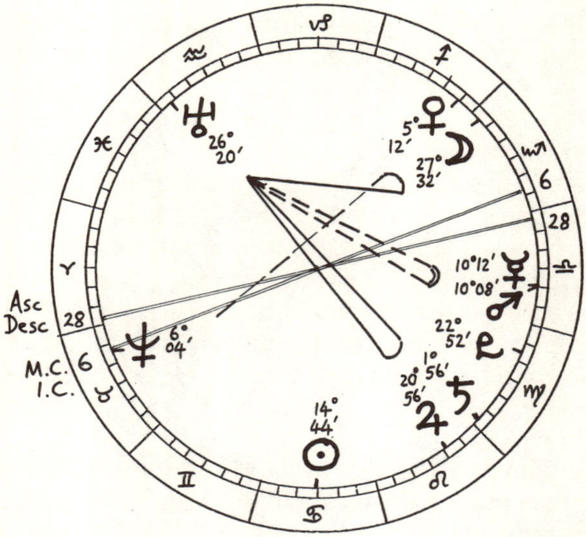

Lucien Van Impe: H4 chart

and Pluto – do not seem to have a strong personal significance unless they are linked in with the more personal planets.) Nevertheless, Van Impe's H4 chart cannot be described as entirely weak, since there are five hard aspects coming into Uranus. It seems that, whatever his trainer may have thought, Van Impe *did* have worries; but these were worries entirely connected with Uranus. It was Uranus which prevented him from realizing the feminine, responsive side of his nature (Moon square Venus in the radical chart), and it was Uranus which prevented him from realizing the potentialities of Mars conjunct Mercury. Just possibly, Uranus represents the trainer himself, who was the 'enemy' with whom Van Impe had to cope.

Conclusion

We can now make suggestions for the interpretation of particular aspects in the H4 chart, based on the examples given in this chapter and on a large number of other cases.

Conjunctions

When two planets are conjunct in the H4 chart, they are square, opposite or conjunct in the radical chart, and so we must refer back to the radical chart for their interpretation. If they were square in the radical chart, they can be regarded as forces which the person is striving to reconcile within his personality.

Oppositions
When two planets are opposite in the H4 chart (and therefore semi-square in the radical chart), the person feels that he lacks the qualities associated with the combination of these planets, and that he must strive to overcome this lack by manifesting these qualities to the world in his external behaviour. Hence he will probably seem to other people to possess these qualities in very large measure, but he himself will never cease to feel his inadequacy and so will be driven on to greater effort.

Squares
When two planets are square in the H4 chart, the person feels that they represent an obstacle which is preventing him from manifesting his qualities to the outside world. Often he will identify himself with one of the planets, and ascribe the qualities of the other planet to his perceived 'enemies' in the outside world. He may fight these 'enemies', and may provoke them into behaviour which makes it appropriate for him to fight; but he may also use their 'hostility' as a reason for inaction and for failing to develop the potentialities of the planet with which he identifies. If a planet is square to two other planets which are in opposition (forming a T-square), this is likely to result in a strong pattern of aggressive manifestation.

Semi-squares
Semi-squares in the H4 chart may not be of great significance, unless they are linked to a pattern of planets that are opposite or square to one another. If they are so linked, the semi-squares represent a further level of obstacle that must be overcome before the obstacles represented by the squares and oppositions can be resolved.

Trines
Trines in the H4 chart may be trines , sextiles, semi-sextiles or quincunxes in the radical chart, and the interpretation must therefore depend on their relationship in the radical chart. If the planets were semi-sextile or quincunx in the radical chart, the person is *struggling to enjoy* the bringing together of these planetary forces.

Sextiles

Here the pattern of 'struggling to enjoy' is still clearer, with perhaps a stronger emphasis on external manifestation: the person feels that he must show the world that he can enjoy behaving in the manner associated with the combination of these planets. These aspects carry connotations of hectic and demonstrative pleasure, but also of unease and embarrassment.

Semi-sextiles and quincunxes

These are probably insignificant if isolated from other aspects. If they are linked into a pattern of squares and trines, they can show how the person links the 'easy' and 'difficult' sides of his H4 pattern of aspects into a single unified pattern of behaviour.

In general, the H4 chart can tell us about the problems that a person feels that he has in dealing with the outside world, with his physical environment and with the people who surround him. It shows the nature of the deficiencies which he feels that he has in dealing with the world, and how he will strive to overcome these deficiencies. It helps to reveal the obstacles which he feels that he faces in the world itself, even though these obstacles may in fact be projections of parts of his own nature. With a strong H4 chart, a person will tend to adopt a fighting stance towards the outside world. With a weak H4 chart, his attitude to the world will be far more acquiescent, or perhaps even indifferent.

CHAPTER 5

THE FIFTH-HARMONIC CHART

When we move on to the fifth-harmonic (H5) chart, we move for the first time into unknown territory. For whereas the H4 chart was concerned with the principle of Twoness which is already familiar to astrologers, the H5 chart is concerned with the unfamiliar principle of Fiveness. We cannot begin to understand the H5 chart until we have grasped something of the essential nature of Fiveness.

We can see straightaway that the principle of Fiveness is likely to take us into uncharted territory. All the traditional astrological aspects are based on combinations of Two and Three, and astrologers have hitherto found them adequate for describing the ways in which the planetary forces are related to one another. Moreover, Geoffrey Dean's researches[1] have shown that when a planet is unaspected (that is, when it has none of the traditional Two and Three aspects with any other planet), the person feels it to be a separate and unintegrated part of his personality. If that person succeeds in forging a link between the unaspected planet and other planets through the principle of Fiveness, he is forging an 'unnatural' link: a link which he feels is not naturally present, but which he is deliberately or artificially creating.

This 'unnaturalness' of the Fiveness principle is supported by the analogy with music (an appropriate one, since the whole concept of harmonics is based on musical theory). Virtually all music is based either on a two-beat rhythm (*tick*-tack) or a three-beat rhythm (*tick*-tack-tack), which are the rhythms that occur most naturally to the human ear; and these two types of rhythm are clearly related to the principle of Twoness (the dynamism and tension of the march rhythm) and of Threeness (the lightness and pleasantness of the waltz rhythm). But it is extremely rare to find music based on a five-beat rhythm (*tick*-tack-tack-tack-tack). If such a piece of music were found, its rhythm would probably strike the listener as 'unnatural', and his ear would take some time to adjust to it. In music, the composer has freedom to link notes together in whatever way he chooses, so that he can always use the 'natural' two-beat or three-beat rhythm and never needs to resort to five beats in a bar. But in astrology this freedom does not exist. If two planets are not naturally linked by a Twoness or Threeness aspect, then we must strive to form a Fiveness link between them.

We can surmise, then, that Fiveness is connected with the *deliberate forging of links* between features that are naturally felt to be separate. A similar idea is conveyed by what numerologists have said about the number Five. Bosman says:

> Five, according to its root meaning, is the number of harvesting, of *arranging* the 'sheaves' of the produced Substance, the hitherto potential substance which now becomes matter . . . . The number five represents that Cosmic process during which Matter is qualified, separated into kinds, and *arranged,* like the harvest, for *use.*[2] (my italics)

Similar conclusions have been reached by astrologers investigating the nature of the fifth harmonic. John Addey says that Five represents 'the putting together of form and matter and in this sense art'.[3] The making of art is, of course, artifice: the creation of artefacts, or things that do not naturally exist. Addey also says: 'Notice that this process is accomplished by mind, for it is mind which can subjectively take into itself the idea of formal principle and the idea of matter and so unite them. Hence, note, a relationship between Five and *mind.*' John Addey and Charles Harvey[4] also point out that Five is connected with *power,* since to be able to arrange matter into a form of one's choosing is to exert power over it.

My own research has convinced me that Fiveness is essentially connected with the idea of *making, arranging, building, constructing, structuring, forming.* It is to do with the creation of order out of chaos: the bringing-together of things that are naturally separate into a formal relationship with one another. It is, therefore, the first number in which man asserts his power over the world. Within the principle of Twoness, he accepts the world as it is and struggles to find his place in it; within the principle of Threeness, he accepts the world as it is and revels in it; but within the principle of Fiveness, he strives to change the world and to make it other than how he found it.

As in the previous chapter, we will look at a number of examples of people with strong H5 charts, so as to try to establish

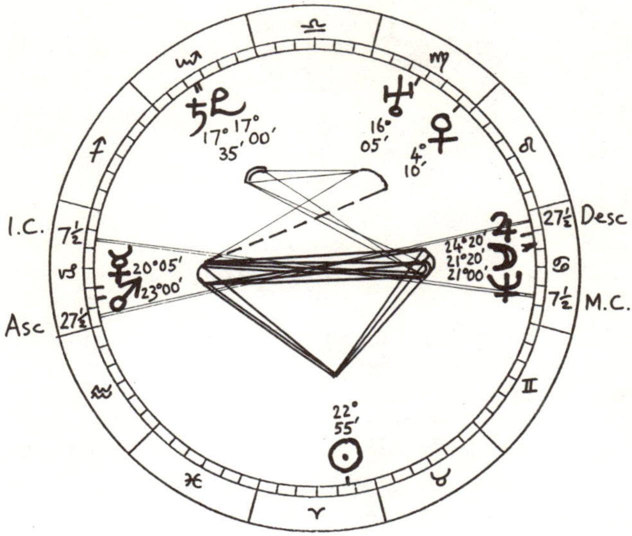

'Thomas': H5 chart

what kinds of activity are particularly associated with the fifth harmonic.

Making things

As we would expect, people whose main preoccupation is with *making things* tend to have strong H5 charts. An outstanding example of this is an engineer, whom we will call **'Thomas'.** I have not met Thomas, but his sister has described him to me as follows:

> In childhood he used to disappear down the garden to the shed and make things – engines, model airplanes, etc. Very 'close' and had only one or two close friends. Never a gang member, never fought physically, never had a girlfriend or belonged to clubs or such-like. Very slow at reading . . . . Poor Thomas, he suffered at school – a posh grammar-type school, when all he wanted was engineering, physics, maths. He is an engineer, now in fact a technical director and personnel manager. He is and always was 'nervous' or highly strung, and very precise. He is intelligent, clings to his basic truths and puts them into practice. He hides his real emotions and suffers physically for it – especially throat and stomach.

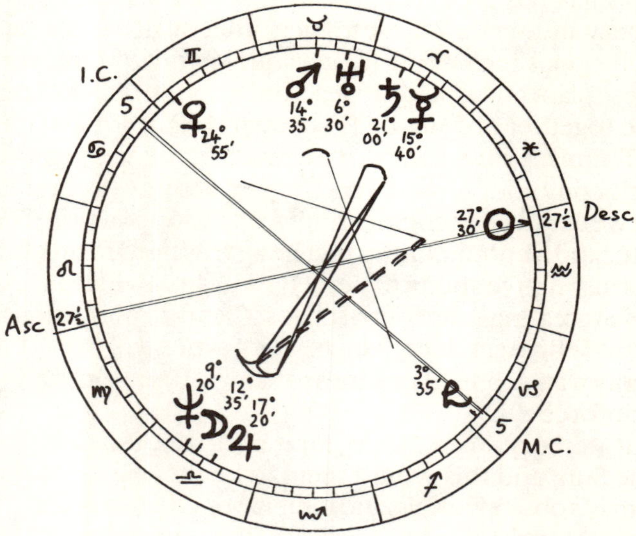

Albert Einstein: H5 chart

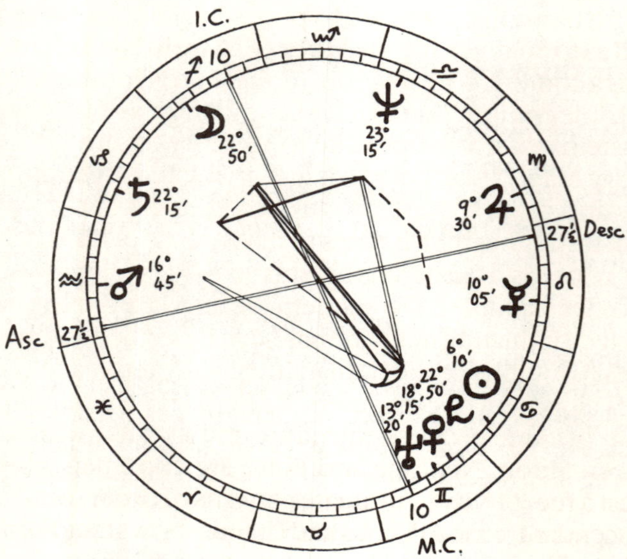

Alan Turing: H5 chart

Thomas has an exceptionally strong H5 chart (see p.49). Mercury and Mars are close to each other on the Ascendant, and closely opposite to Moon, Jupiter and Neptune. (Moreover, none of these planets is aspected to each other in the radical chart; they come together only in the H5 chart.) Mercury and Mars show that Thomas has a strong drive towards *making things* in a way which combines physical and mental activity: a combination very well suited for engineering. And the opposition to Moon, Jupiter and Neptune shows that he is striving to relate this drive to his instinctive ability to perceive imaginative and expansive ways of creating order out of chaos. Clearly Thomas is not only a natural engineer or craftsman, but a potentially very creative and innovative one. (The trines to Saturn, Uranus and Pluto help to reinforce this.)

But Mercury, Mars, Moon, Jupiter and Neptune are all square to the Sun, and this would seem to be the source of much of Thomas's obsessive traits and inhibitions. This is a point to which we will return later, when we deal with the relationship between H5 and the obsessive personality.

Science and mathematics

The fifth harmonic is related to pure science and mathematics as well as to engineering. In **Albert Einstein's** H5 chart (p.50) there is a conjunction of Moon, Jupiter and Neptune: this is the same triple conjunction that we saw in Thomas's H5 chart, but in Einstein's case it is opposite to Mercury-Saturn rather than Mercury-Mars. Thus Einstein is less orientated towards making things and more towards controlled thought and communication, through which he tries to make sense of his receptivity to the patterns of order in the universe.

Another example is the mathematician **Alan Turing,** whose language-simulating machine has been described as 'one of the greatest triumphs of the human intellect'[6] and made possible the development of computers. Turing's H5 chart (p.50) has a strong cluster of planets centred around the Moon: the aspects between Moon, Neptune and Pluto are extremely close and suggest a receptivity to ultimate patterns of order. But Mars on the Ascendant, trine to Venus and Uranus shows that Turing was a maker of objects as well as a pure scientist.

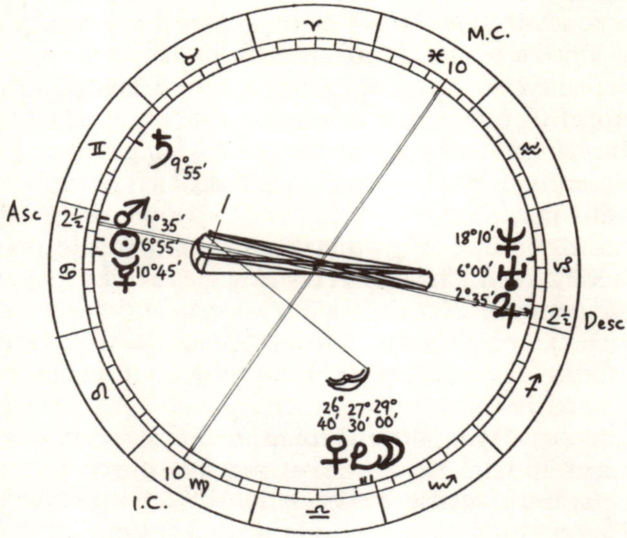

W.A. Mozart: H5 chart

The creation of forms in art

The fifth harmonic can also relate to the creation of forms and structures in art. The composer **Mozart** has an exceptionally strong H5 chart, with five planets tightly clustered on the Ascendant-Descendant axis. Mozart was a compulsive creator of musical forms. Scholes has a picture of 'Mozart at the Billiard Table'. and comments:

> Ball games, particularly billiards and bowls, were greatly to his liking. There is little doubt that he pursued these games not merely for their own sake but because he found in the movement and control of a rolling ball a congenial accompaniment to the movement within his own copious and productive mind.
>
> Instances are recorded of his stopping in the middle of a game to make notes, or of his humming, as he played, a theme which was later found in one of his works. Moreover, he was particularly fond of playing billiards alone, keeping his note-book handy – though the notes he made were always the briefest indication of an idea, for he did his actual composing in his head.
>
> The ever-flowing rhythms in his mind induced him incessantly to tap his fob, a table, a chair-back, or anything to hand, and there is no doubt that he spent some of his most fruitful hours alone at the billiard table.[6]

Mozart's H5 chart (p.52) is far stronger than any of his other harmonic charts, and he is perhaps an extreme example of a person who lived naturally and continually at the fifth-harmonic level of reality, which is concerned with the creation of order and pattern. Compulsive musical composition and compulsive billiard-playing were two manifestations of this, which he was able to carry out simultaneously. The fact that he lived continually within his private world of Fiveness helps to explain his naivety and childlikeness within the ordinary world of Twoness and Threeness which most of us inhabit most of the time.

In Mozart's H5 chart, the Sun-Mercury and Moon-Pluto conjunctions are carried over from the radical chart. Thus, the most striking features of the H5 chart are: firstly, the conjunction of Sun-Mercury with Mars on the Ascendant, opposite to Jupiter and Uranus; and secondly, the conjunction of Moon-Pluto with Venus, trine to Mars. Mars is the link between these two clusters of planets, and it seems to be Mars in H5 which above all drives people to create things which have the H5 characteristics of order and pattern and structure and form. The first cluster of planets (Sun-Mercury-Mars-Jupiter-Uranus) is similar to the pattern which we saw in Thomas's chart, and we can perhaps deduce that Mozart could, if he wished, have been a brilliant engineer or craftsman. But it is the Moon-Venus conjunction – translatable, perhaps, as receptivity to beauty of form – which makes it more likely that Mozart would be a brilliant musician than a brilliant engineer. The involvement of Pluto gives this conjunction a relentless, hyperactive quality, and the trine to Mars means that Mozart is easily able to translate his formal conceptions into actual compositions.

Painters, too, tend to have strong H5 charts, especially those painters whose main concern is with structure and form and order. **Georges Braque,** the Cubist painter, is described by Hughes as 'one of the arch-classicists of modern painting'.[7] He was concerned all his life with the development and perfection of painting style and texture: 'Braque skimped nothing in his quest for a perfectly clear texture of thought'. His H5 chart (p.54) has an *exact* conjunction of Sun and Mars (coming together with the radical conjunction of Sun, Saturn and Neptune) opposite to the Moon, and also a very close Mercury-Uranus opposition linked to the Moon by semi-squares: the perfect combination for someone who was able to find complete self-fulfilment through the creation of disciplined forms, and who also had radical and striking ideas about form and structure which he was able to translate into actuality.

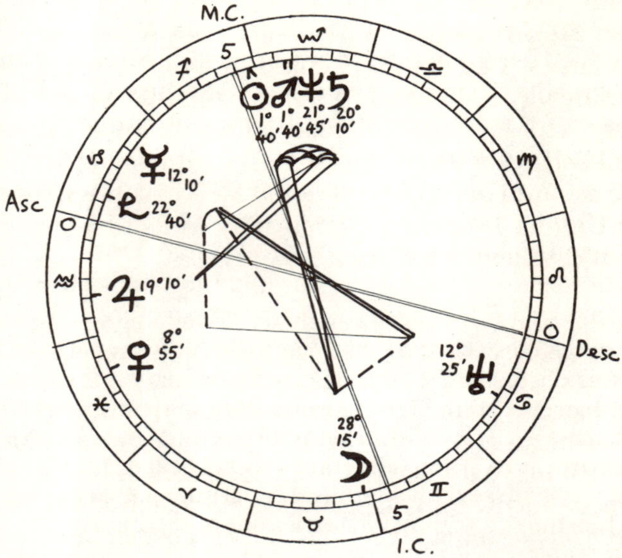

Georges Braque: H5 chart

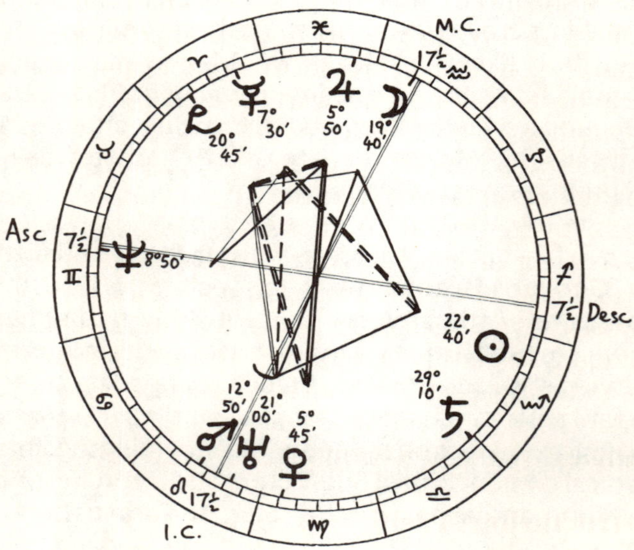

Ernest Hemingway: H5 chart

Another instructive example is the H5 chart of the writer **Ernest Hemingway.** We would expect Mercury to be particularly strong in the H5 charts of writers; but, if we look at Hemingway's H5 chart (p.54), we see that (apart from the Mercury-Neptune sextile) Mercury is linked to other planets only by semi-squares, with an exact semi-square to the sun and a semi-square also to Uranus (thus linking Mercury in with the Sun-Moon-Uranus T-square). Hemingway in fact had very great difficulties as a writer. Raban says:

> In his life-long rage to find the right words Hemingway exercised the desperate craft of a fisherman adrift on stormy water . . . . He was no lord of the language; his vocabulary was sparse, his grammar even thinner . . . . Again and again Hemingway writes of language as if it were another medium. Sometimes he compares it to paint, sometimes to musical notes. What he registers in these similes is the essential intractability of words. For him they were difficult, unyielding things, more like wood than paint, and he laboriously whittled and gouged at the language in the effort to shape it to his will. He fashioned a style out of his inarticulacy which is capable of saying a few things better than any writer has ever said them before.[8]

The semi-squares to Mercury in the H5 chart seem to be an indication of this struggle to find an appropriate style of communication. They represent the combination of Fiveness (the search for order and form) with Twoness (the determination to overcome a lack or a deficiency).

In all these cases of creative people with strong H5 charts, the fifth harmonic seems to refer to the type of *form* and *order* that the artist is trying to create. In other words, it relates to *style* rather than to *content,* or to the classical rather than the romantic mode of creativity. Mozart is the prime example of the classical type of composer, whose genius lies primarily in the creation of form; and Braque, despite his modernity, is a classical type of painter. Artists and composers for whom *content* and *message* are more important than *form* and *style* will tend to be stronger in the seventh than in the fifth harmonic.

The structuring of self

So far, we have considered the fifth harmonic in connection with the creation of forms and structures *outside oneself,* whether these are concrete objects, paintings, musical compositions,

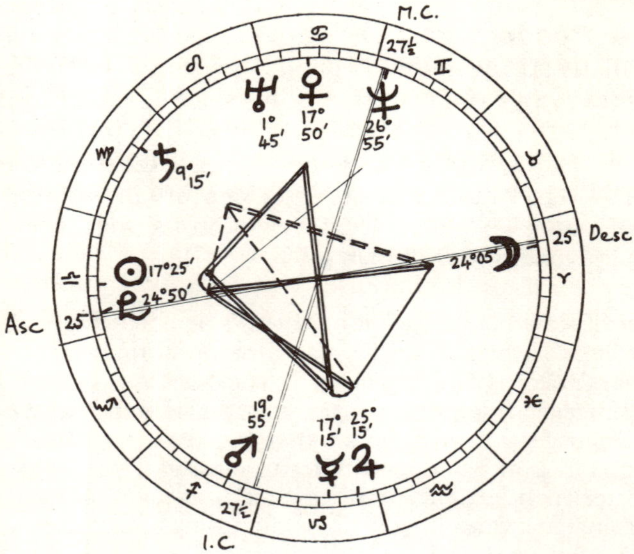

Barry Humphries: H5 chart

scientific theories, or writings. But the fifth harmonic is also concerned wth the *structuring of oneself,* or of one's own behaviour. In fact this, it seems to me, is likely to be closer to its root meaning, and the creation of objects or works of art is only an externalization of what starts as an internal process. The H5 chart is a clue to the type of order and style which a person will attempt to impose upon his own life and his own day-to-day behaviour. The most important key-word here seems to be 'style'. A person with a strong and well-integrated H5 chart will tend to be a more stylized – perhaps even a more stylish – person than one with a weak H5 chart.

The fifth harmonic therefore has about it a ring of artificiality (making a 'work of art' out of one's own life). An unstylized person is one who is easily able to live life in accordance with the 'natural' principles of Twoness and Threeness. A person governed by Fiveness is one who seems to have abandoned natural and spontaneous expression in favour of a deliberately constructed order and style. This style can be of any type, depending on the nature of the planets that are dominant in the H5 chart. It can, for example, be a tightly-disciplined Saturnian style, or a Jupiterian style of flamboyance and exuberance.

Actors

The fifth harmonic therefore tends to be strong in the charts of actors, or at least of those actors whose art consists of 'turning themselves into somebody else', moulding their behaviour in order to create a style. A remarkable example of this is **Barry Humphries**, the actor who is best known for his impersonation of 'Dame Edna Everage, Superstar'. Humphries's H5 chart (p.56) is exceptionally strong and is, if not a portrait of Dame Edna, at least a portrait of the impulses which drive Humphries into identifying himself with an outrageous personality of the opposite sex.

Firstly, there is the very close T-square of Sun, Mercury and Venus. (This T-square cannot of course occur in the radical chart, since Sun is never square to Mercury or Venus; but in the harmonic charts it can occur, and is of very great significance.) The T-square seems to say: 'I must strive to find a style through which I can communicate with the world so as to attract people and win their admiration.' Secondly, there is another very close T-square of Moon, Pluto and Jupiter (with Pluto conjunct Sun on the Ascendant), which is an indication of the extreme exuberance of Humphries's style and the Plutonian relentlessness of his mockery. Thirdly, Saturn is semi-square to Moon, Jupiter and Pluto, which perhaps shows Humphries's efforts to control the exuberance of his style so that he can repeat the same performance night after night in the theatre. And fourthly, Neptune on the Midheaven perhaps represents the fact that, through his performance, Humphries seeks glory and adulation. Mars is unaspected, which perhaps helps Humphries to impersonate a woman rather than a man; but the lack of Martian dynamism is more than compensated for by Pluto, which is the pivotal planet of the whole chart. Humphries is in fact an example of a person who seems to have been taken over by his H5 chart, so that he is uncertain whether he is himself, or the artificial character which he has created. When he appears in public as himself, he seems uneasy and lacking in confidence: he only finds self-confidence behind the mask of a style.

Humphries's acting style, despite its humour, is very tense and abrasive, and this may be related to the fact that his H5 chart is dominated by 'hard' aspects (oppositions, squares and semi-squares). If an actor's H5 chart is dominated by 'soft' aspects (trines and sextiles), it is likely that his acting style will be more relaxed and easy-going. An example of this is **Liberace,** the

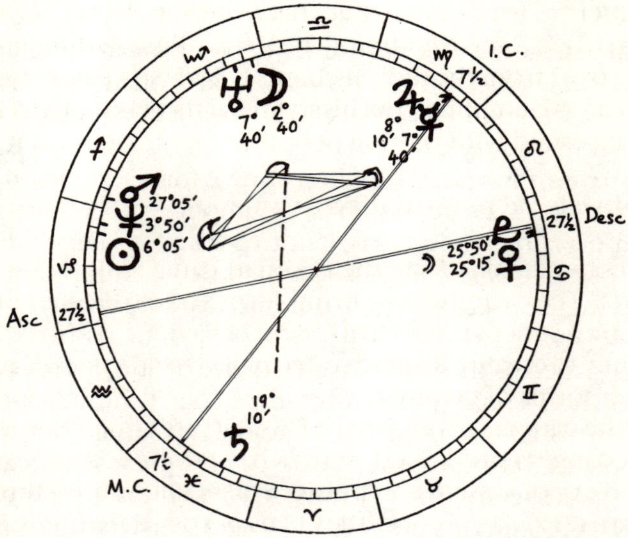

Liberace: H5 chart

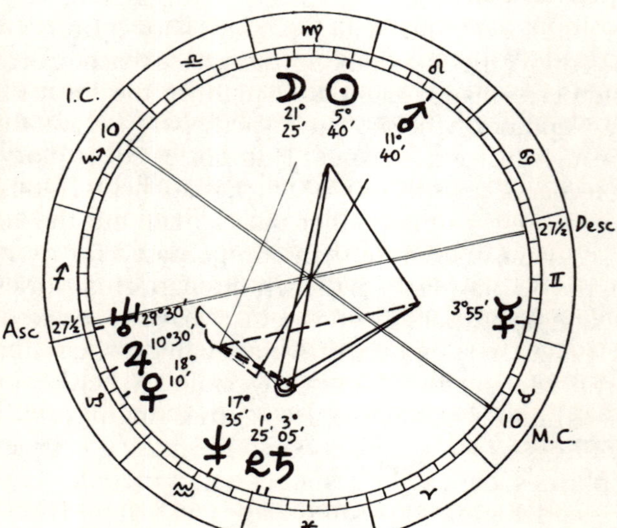

Jean Cocteau: H5 chart

world's highest-paid pianist, whose H5 chart (p.58) is dominated by a pattern of trines and sextiles between Sun, Moon, Mercury, Jupiter, Uranus and Neptune. (The Venus-Pluto conjunction is carried over from the radical chart.) Liberace is just as much a showman as Humphries, but his style is far more cool and relaxed and concerned with putting people at their ease.

The obsessive personality

Links between planets in the H5 chart can clearly bring great benefits. They can help to bring form and order into life, to develop a distinctive, personal style of behaviour, and to develop the ability to create form and order in the world through artistic or other forms of creativity.

But the dangers of links in the H5 chart are equally apparent. These dangers are linked with what psychologists call the obsessive or obsessive-compulsive personality. Fifth-harmonic links can create a style, but they can also make that style into a prison. Storr says: 'Perhaps the most striking feature of the obsessional temperament is the compulsive need to control both the self and the environment. Disorder and spontaneity must be avoided so far as is possible, since both appear threatening and unpredictable.'[9]

Oppositions and squares in the H5 chart seem particularly likely to lead to obsessive or compulsive behaviour. We have seen signs of this already in the H5 chart of Thomas the engineer, where it seems likely that the squares between five planets and the Sun were at least partly responsible for his mental and emotional inhibitions. Mozart's compulsive billiard-playing can also be seen as an example of obsessive behaviour. Mozart was able to use this as an aid towards creativity, but for other people, who lack a creative outlet, such repetitive behaviour patterns can easily become simply an escape into an artificial world of form and order which bears no relation to the chaos of the real world. In some cases these obsessive patterns can be a hindrance, rather than a help, to creativity. **Jean Cocteau,** the poet, artist and film-maker, said:

> My worst defect comes to me from childhood, like nearly all that I have. For I remain the victim of those obsessive rhythms which make some children idiots . . . arranging their plates in a special way at table, or stepping over certain grooves in the pavement. In the midst of my work these symptoms grip me, forcing me

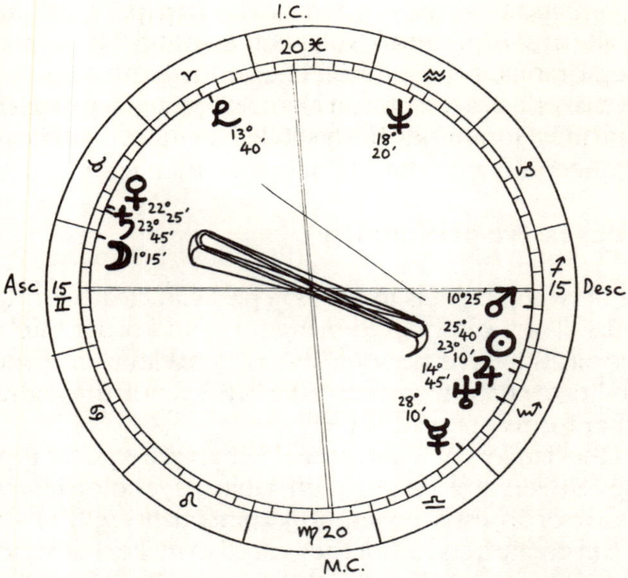

'Sick Child': H5 child

> to resist that which is pushing me along, involving me in some strange crippled style of writing, hindering me from saying what I would.[10]

It is easy to see the origins of this obsessive behaviour in the Sun-Mercury-Saturn-Pluto T-square in Cocteau's H5 chart (p.58). Note the reference to '*arranging* the plates in a special way', which takes us back to Bosman's reference to 'arranging' as part of the essential meaning of the number Five. The fifth harmonic is connected with the instinct to *arrange* the parts of both the self and the environment into an order which is felt to be satisfactory. Obsessiveness seems to come when the 'arranging' is carried out *for its own sake,* without reference to any *use* to which the arrangement can be put.

Our final case is one of self-destructive behaviour in childhood, which can possibly be attributed to obsessive attitudes resulting from a strong H5 pattern. This is the case of the **sick child** described by Ebertin in *The Contact Cosmogram.*[11] Ebertin says of this child:

> In the first two years of his life he vomited daily, and sometimes even at every meal. He contracted measles at the age of eighteen

months, mumps at five years of age, chicken pox at six, and in the late summer of 1960 the child's legs were covered with blotches of varying sizes. The aftermath of the chicken pox saw the child gradually deteriorating, with complete loss of appetite; weekly blood transfusions became necessary, and yet he could not be saved.

In the child's radical chart, Venus and Saturn are in very close conjunction in the sixth house (Placidus), but in the H5 chart (p.60) Venus and Saturn come into conjunction with the Moon and also into close opposition with Sun and Jupiter (which are also in conjunction with Uranus). On the principle that many illnesses have psychological causes, it seems at least possible that this strong H5 pattern contributed to the child's ill health and therefore to his early death. There are indications of a type of obsessive behaviour in which the child, acting from within his instinctive Moon-personality, would constantly strive to thwart and repress his own Sun-Jupiter-Uranus vitality. This could well cause his rejection of food and his lack of resistance to infection. But if the child could have lived beyond this instinctive phase to an age at which he could have identified more with Sun and Jupiter, then the positive and creative potentialities of the chart might eventually have become apparent.

Conclusion

As a result of these and other cases, we can now make suggestions for the interpretation of aspects in the H5 chart.

Conjunctions

When two planets are conjunct in the H5 chart, the person has the desire and the ability to bring them together in an ordered, patterned and structured way. This may manifest in his own behaviour and his presentation of himself: he may develop a style of behaviour in which the characteristics of the two planets are combined. This style may well seem to be artificial, in the sense that it is different from his normal or unstylized mode of behaviour in which he reacts spontaneously to his environment. Or it may manifest in the type of order and structure which he seeks to impose on his environment: in this case it may result in the creation of artefacts which display the characteristics of the planets in their form and style.

Oppositions

When two planets are opposite in the H5 chart, the person is *seeking* to bring them together in an ordered and structured way. The interpretation is the same as above, except that there is more of a struggle, a determination to overcome a lack, or a restless search for an appropriate style. Sometimes the person may identify with the planet(s) at one end of the opposition, and develop a pattern in which the planet(s) at the other end of the opposition are repressed or thwarted.

Squares

Here there is still more of a struggle: there will be great expenditure of effort towards finding an appropriate style. While this may result in great creative achievement, it is likely that the resulting style will contain a harsh, abrasive or aggressive quality (though this will of course be affected by the nature of the planets concerned). There is a danger that the person may become locked in obsessive patterns in which he indulges in repetitive and purposeless activities: the effort to find a style which accommodates the two planets has become an end in itself.

Semi-squares

These may be insignificant in themselves, unless they are linked in with two or more planets which are opposite or square to each other. When they are so linked, the planet which is semi-square to the other planets may represent a faculty which the person feels is missing from the style which he is struggling to develop, and which he feels that he must integrate into that style. He never ceases to feel that his style lacks the qualities of this planet; he also never ceases in his efforts to include it.

Trines

When two planets are trine to each other in the H5 chart, the person finds pleasure and enjoyment in linking them together in an appropriate structure or style. The result is likely to be a style which contains the qualities of lightness, relaxation and fluidity. As with the other aspects, this may manifest either in the person's own behaviour or in the artefacts which he creates. When the person engages in creative activity, there is likely to be a great natural facility in the activity, but a relative lack of awareness of inadequacies and hence a lack of ability to change and improve one's style in response to pressure.

Sextiles
When two planets are sextile to each other in the H5 chart, the person is *seeking* to find the type of pleasurable style which is associated with the trine.

Semi-sextiles and quincunxes
These are probably insignificant in themselves, except when they are features of a larger pattern and show how the person can link together the 'difficult' and 'easy' features of his H5 chart into a single unified style.

We can conclude this chapter by suggesting that Five can perhaps be seen as man's own number. With his five fingers on each hand and his decimal system of counting, man has naturally seen Five as the number which helps him to make sense of the world and claim it as his own; and we have seen that Fiveness is the principle which is concerned with the creation of order and form out of disorder and chaos. Man seeks continually to classify and categorize things in order to understand them; and he seeks also to mould the world about him into forms and structures of his own making, and to mould his own behaviour into a style that suits his purposes. Man's ability to use the principles of Twoness and Threeness may not be much greater than that of the other animals, but his ability to use the principle of Fiveness is greatly superior. Fiveness is thus the prime source of man's creativity, and of his capacity for intellectual understanding. But it carries with it the inherent danger that, by moulding the world into artificial forms and structures, we may lose touch with the world as it really is; and by moulding our own behaviour into an artificial style, we may lose touch with ourselves as we really are.

CHAPTER 6
THE SEVENTH-HARMONIC CHART

Seven is the next prime number after Five, and the last prime number before Eleven. Seven, like Five, takes us into uncharted territory, beyond the reach of the Twoness and Threeness in terms of which astrologers have traditionally interpreted charts. If two planets are linked together by the principle of Sevenness, they will relate to each other in a way which is different from the Twoness principle of effort and challenge, the Threeness principle of pleasure, or the Fiveness principle of order and form.

Seven has traditionally been seen as a magical and mysterious number. Five (as we said at the end of the previous chapter) can be seen as the number of man: but Seven is seen as a number of God: a number which contains cosmic secrets and divine insights. Human order may be found in the number Five, but cosmic order is sought in the number Seven.

However, it seems unlikely that the universe is patterned according to the number Seven. (Man has attributed the seven days of the week to divine guidance; but in fact they are more likely to be a human invention. Man thought that there were seven planets, and built a whole cosmology around this belief; but in fact there are more than seven.) Seven seems to relate more to man's fantasies about the universe than to its objective reality.

Charles Harvey[1] has said that the seventh harmonic reveals what aspects of life will inspire us and 'turn us on'; and this is, I think, close to the essential meaning of the seventh harmonic. However, Harvey goes on to say that the seventh harmonic reveals what will 'lift us into a higher order of vision and understanding. It shows our receptivity to the influx of supernal energies and circumstances which can thus open us to the "breathing in" of the Divine.' In my view, this divine insight and understanding is closer to the meaning of the ninth harmonic (which we shall discuss in the next chapter) than to that of the seventh.

Charles Graham[2] has written a valuable paper on the H7 charts of creative artists; but, since he is entirely concerned with the relationship of the seventh harmonic to creativity, he does not consider whether it has any wider meaning.

In this chapter I shall propose that Seven is in fact a number of man: a number not of man's rational and constructive abilities, but of his wild, fertile and unpredictable imagination. It is a number of inspiration, but the inspiration proceeds from within man himself. It is a number of man's highest flights of creative imagination, and of his lowest depths of destructive illusion. It is a number, not of permanence, but of transience; not of unchanging truth, but of sudden flashes of light and darkness; not of objective reality, but of subjective impressions and the emotional experience which man derives from those impressions; not of knowledge, but of fantasy. It is, in short, a romantic number. People[3] have struggled, without much success, to define the word 'romantic'; but, even without a definition, its meaning is well understood, and is close to the essential meaning of the number Seven.

When two planets are linked in the seventh harmonic, the person is aware of a *romantic* link between them. He cannot bring them together through the natural principles of Twoness and Threeness, nor even through the more artificial principle of Fiveness; and yet he longs to bring them together, he is inspired by the idea of bringing them together; and this idea – of the combination of the energies of these two planets – will have for him a strong emotional power. In his imagination (whether consciously or subconsciously) he may construct fantasies around this idea; and these fantasies affect his view of the world, and so may come to have an effect upon his visible behaviour. These manifestations may take many forms: he may feel that he has a message to convey to mankind, and may attempt to convey it through creative art; or he may have a romantic vision of the future of mankind, and attempt to convert this vision into reality by fighting for a cause or through political action; or his fantasies may be purely personal, concerned only with his perception of himself and of other people in relation to himself, and so affecting his day-to-day actions and interactions. (For instance, squares and oppositions in the H7 chart may mean that the person is inspired by the idea of *conflict between* these planetary forces, and so may come to feel that this conflict is in fact taking place between himself and the people around him.)

In this chapter we will look at a number of examples of strong and well-integrated H7 charts. But the seventh harmonic

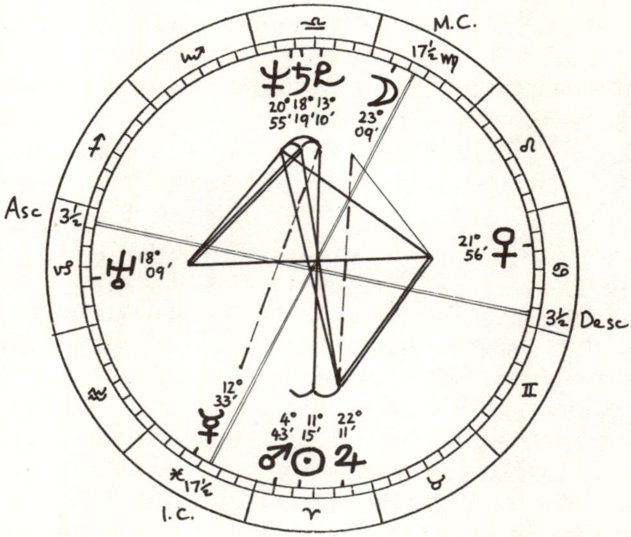

Hector Berlioz: H7 chart

contains tremendous potential for both good and bad. We will start by looking at some people who have put the creative potential of the seventh harmonic to good use, and we will move from there to the 'bad' end of the spectrum.

Creative artists

Creative artists are certainly not the only people with strong H7 charts, but they are among the people who seem to bring the seventh harmonic out into the light of day, whereas for many other people it remains hidden beneath the surface of their lives. We will therefore examine several examples of the H7 charts of creative people.

We can start with the composer **Hector Berlioz.** Berlioz's H7 chart is extremely strong (see p.66): it has a Grand Cross involving *all five* of the outer planets and also Venus (with the Sun also involved by being conjunct Jupiter and opposite Pluto), and the squares from Venus to Jupiter and from Saturn to Uranus are extremely close. The involvement of Venus is important because it is Venus in the H7 chart which above all represents romanticism in the popular sense, and especially romantic love.

Berlioz is the archteype of the romantic composer (just as

Mozart, whose main strength is in the fifth harmonic, is the archetype of the classical composer). Scholes says:

> The romanticism of Berlioz was innate. His romantic sensibility showed itself in childhood when he would weep at a touching phrase of Virgil, and it expressed itself all through his life in a series of love affairs, varied and numerous even for his nationality and his period . . . . Another romantic characteristic of his was his love of immensity . . . . Another (and very marked) characteristic is the literary influence, his works often being the musical re-expression of the romantic poets . . . . Finally may be mentioned the romantic characteristic of 'programme' in instrumental music, i.e. a scheme embodying in tone emotions aroused by a series of events or by visible scenes.[4]

Moreover, Scholes, having described a number of Berlioz's romantic love affairs, goes on to say:

> These romances . . . . may seem irrelevant in a discussion of his position as a composer, but really are directly to the point, as life and art were one with him and such incidents show him as true to one principle of the extremist romantic theory – *not Rule but direct reaction to Feeling.*

This is an excellent description of the H7 mode of behaviour (as opposed to the H5 mode which is concerned with Rule), and it is clear that Berlioz, with his Grand Cross in the H7 chart, strove to live his life according to romantic ideals and to express those ideals in his music. But also, his musical inspiration was intimately bound up with the emotional tribulations in his non-musical life.

The planet which is least integrated in Berlioz's H7 chart is Mercury: but Mercury is very well integrated in his *fifth*-harmonic chart, where it is exactly conjunct Saturn, trine to the Moon and square to Venus, and clearly he used these H5 abilities to impose order and form on his music. Most great artists seem to need strength *both* in the fifth *and* in the seventh harmonic, and both of these harmonics (as well as the ninth, which we will come to in the next chapter) need to be considered in order to understand the nature of their art.

(Berlioz is also the composer of the only musical piece I know which has a rhythm of seven beats to a bar. This is the 'Dance of the Soothsayers' in Berlioz's *Childhood of Christ,* and its wild and whirling rhythm well suggests the nature of the seventh harmonic.)

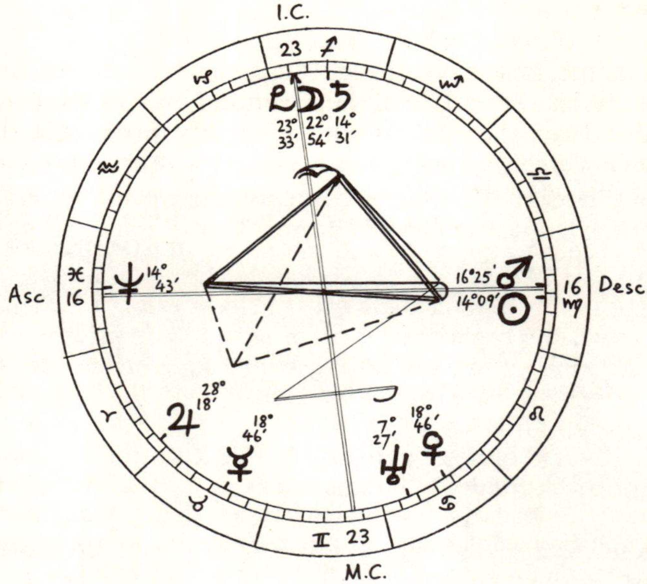

James Ensor: H7 chart

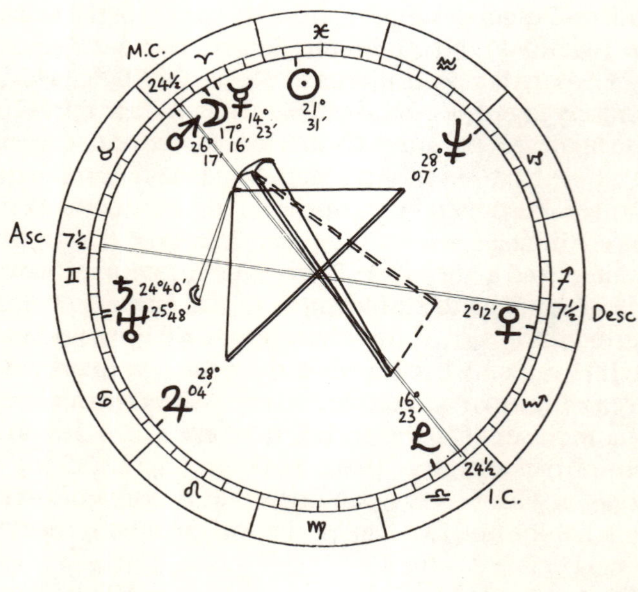

Salvador Dali: H7 chart

Painters

Many painters also have strong H7 charts, which are descriptive not so much of the *style* of their painting as of its *imaginative content*. Thus, the painters who exemplify this most strongly are not necessarily the greatest painters but those whose work shows a strong and consistent content of imaginative inspiration.

As an example, we can take **James Ensor,** a Belgian painter who worked in isolation and belonged to no school, but is recognized as the greatest Belgian painter of the late nineteenth and early twentieth centuries. Ensor was obsessed with painting skeletons and masks. The following are some extracts from the description of Ensor's work in the *Phaidon Dictionary of Twentieth-Century Art*:

> Sense of the grotesque . . . sour humour . . . . It seems that the paranoia induced by the critical attacks upon him combined with his unhappy childhood to inspire the bitterness and horror of much of his painting . . . . He identified himself with Christ: he too was mocked and persecuted.[5]

Ensor's H7 chart shows an extremely strong pattern (see p.00). Sun and Mars are very closely opposite to Neptune on the Ascendant-Descendant axis, and also very closely square to Saturn (with Sun, Neptune and Saturn also semi-square to Jupiter). Saturn is also conjunct to Moon and Pluto which are very closely together on the I.C. Saturn, which in the H7 chart often seems to be the planet of suffering, is thus the pivotal planet of the whole chart. In the Sun-Mars-Neptune-Saturn T-square, we can see Ensor reaching out towards transcendence ('he identified himself with Christ') and feeling thwarted and persecuted, and in the Moon-Pluto-Saturn conjunction we can see the grimness and ruthlessness of the fantasies which he converted into paint. If this chart was in H4, it would show a man who fought his battles out in the world, perhaps with the aim of earning a martyr's crown (whereas in fact Ensor lived the life of a recluse); if it was in H5, it would show a man who presented himself in a harsh, abrasive and shocking *style* (whereas in fact Ensor's painting style is conventional and unremarkable); but, as it is in H7, its content is imaginative and fantastical.

Salvador Dali also has a strong H7 chart (p.68). Its structure is remarkably similar to Ensor's: in place of Moon-Saturn-Pluto

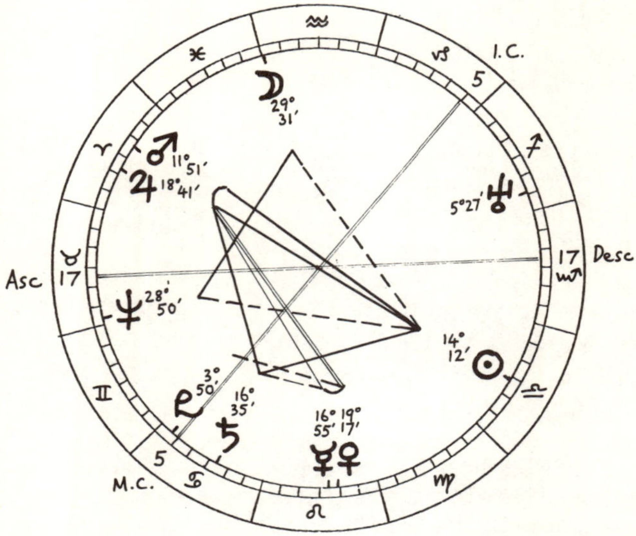

Rupert Brooke: H7 chart

on the I.C., we have Pluto on the I.C. opposite to Moon-Mercury-Mars, and in place of Saturn square to Sun-Mars-Neptune (and semi-square to Jupiter) we have Mars square to Jupiter and Neptune (and also sextile to Saturn and Uranus). But in Dali's case Mars is the pivotal planet, which fits in with the sexual nature of many of Dali's fantasies. Hughes[6] says that Dali 'employed an imagery of impotence and guilt', and the *Phaidon Dictionary*[7] says that some of Dali's paintings 'are totally convincing as visual messages from the human subconscious mind'. But the peculiar brilliance of Dali's painting style is more due to his H5 chart than to his H7 chart (in H5 he has Sun trine Moon, and also Mercury closely conjunct Uranus and opposite to Jupiter).

Writers

An example of a poet with a strong H7 chart is **Rupert Brooke.** In Brooke's H7 chart Sun is opposite to Mars and Jupiter (with Jupiter trine to Mercury and Venus), and Brooke's poems show an imaginative striving towards Martian vigour and Jupiterian vitality expressed in a charming (Venusian) way, even though in 'real life' he found it difficult to express these qualities. In a poem entitled 'Vanitas' written at the age of eighteen, he expresses this as follows:

Laugh now and live! Our blood is young, our hearts are high,
Fragrant of life, aflame with roses, all the Spring
Thrills in our windy souls and woos to wayfaring;
And the glad sun goes laughing up the eastern sky.[8]

But Sun and Jupiter are also square to Saturn, the planet (in H7) of suffering and death; and so Brooke is also imaginatively obsessed with death – not only physical death, but also the death of youth, the death of poetry and idealism, the death of vitality, the death of love. Thus in the third verse of the same poem he goes on:

Only – remember! The day passeth; not for long
Stays the mad joyance of our golden revelry.
The young days darken; the rose petals fade and die;
Sleep ends and crowns our carnival, silence our song.

This theme, crudely expressed in this early poem, recurs over and over again throughout Brooke's work, as for instance in his famous poem 'Day that I have Loved'. In his last poems, inspired by the First World War, this theme became transmuted into the glorification of death in battle ('Blow out, you bugles, over the rich Dead!'), and it seems at least possible that Brooke's own death (which occurred during war service – not in battle but from a combination of illnesses) was hastened by his imaginative striving towards an idealized death.

The seventh harmonic is also related to sexual fantasy and eroticism, especially if Mars is prominent. We have already seen this in the case of Salvador Dali. Another example is **Henry Miller,** whose writings are well known for their erotic content. Miller's H7 chart (p.72) contains a quintuple conjunction of Sun, Moon, Mars, Neptune and Pluto, with Mars *exactly* conjunct Pluto, which well conveys Miller's imaginative preoccupation with his own sexuality. Also Mars, Neptune and Pluto are opposite to Venus. Miller is said to have been frightened of women,[9] and this opposition conveys the idea that he was inspired by the thought of the 'battle between the sexes'.

An architect with a vision

A quite different type of creative inspiration is shown in the H7 chart of **Le Corbusier** (p.72), the architect who, more than anyone else, was responsible for the modern fashion of housing

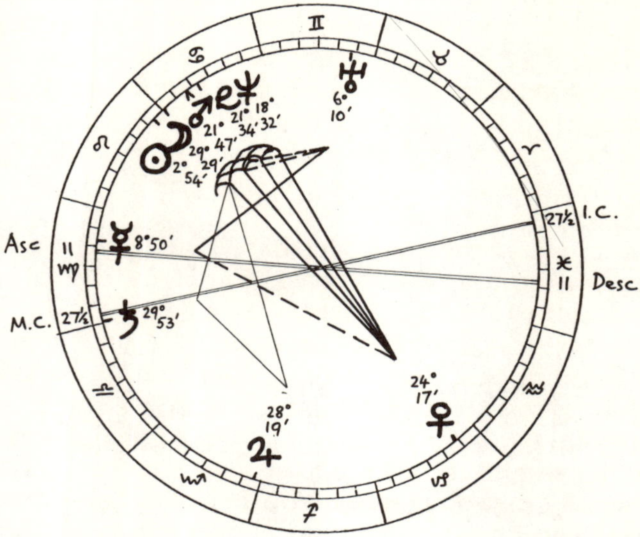

Henry Miller: H7 chart

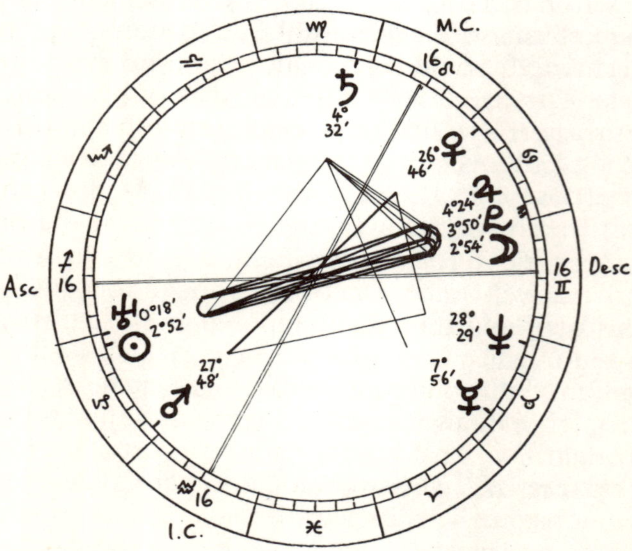

Le Corbusier: H7 chart

people in immense tower blocks ('high-rise mass housing') surrounded by empty green space, and can therefore be said to be responsible for much of the misery of urban living in the twentieth century. Le Corbusier was inspired by the idea of creating a Utopia, by clearing away all the chaos and confusion of the existing city and replacing it by structures in which people would lead the ordered lives of ants or termites. Surveying the confusion of central Paris, he cried: 'Imagine all this junk, which till now has lain spread out over the soil like a dry crust, cleaned off and carted away and replaced by immense clear crystals of glass, rising to a height of over 600 feet!'[10]

In order to understand Le Corbusier's H7 chart, we must note that both the Sun-Uranus conjunction and the Moon-Pluto conjunction (sextile to Saturn) are carried over from the radical chart. But it is in the H7 chart that these conjunctions come together with each other (and also with Jupiter) to form an immensely powerful opposition between Sun-Uranus and Moon-Pluto-Jupiter, with Saturn mediating between them (and also with Mercury trine to Saturn).

This chart seems to show that Le Corbusier's infliction of Utopia on the rest of humanity is the result of his own internal problems and his fantasies about himself. Le Corbusier was a man with a strong need for theatrical self-advertisement (Sun closely conjunct Uranus in the radical fifth house) and also a strong need for seclusion (Moon closely conjunct Pluto in the radical twelfth house). These two very different sides of his nature come together in H7, so that he is seeking an *inspirational* way of bringing them together: an inspirational way of avoiding people through assertion of his own individuality. But how can this be done? The answer lies in Saturn, which links all these planets together in a joyful pattern of trines and sextiles. In this context Saturn ceases to be the planet of suffering and becomes instead the planet of limitations gladly imposed and willingly accepted. It is as though Le Corbusier was saying: 'I will enclose these people in boxes, which will release me from the need to relate to them, and will also cause them to revere me for my brilliance and originality'. That was the nature (or part of the nature) of Le Corbusier's H7 fantasy. (In addition there is the Venus-Mars-Neptune pattern, which makes the picture still more complex.)

Of course it may be that Le Corbusier never consciously thought in this way. Unlike Berlioz, Ensor or Dali, he was not consciously trying to explore his subconscious mind and bring it to the surface; he was working at the H5 level, creating forms

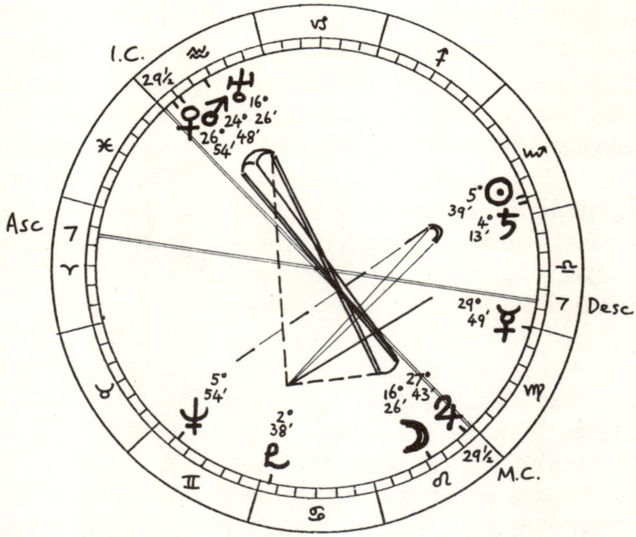

Adolf Hitler: H7 chart

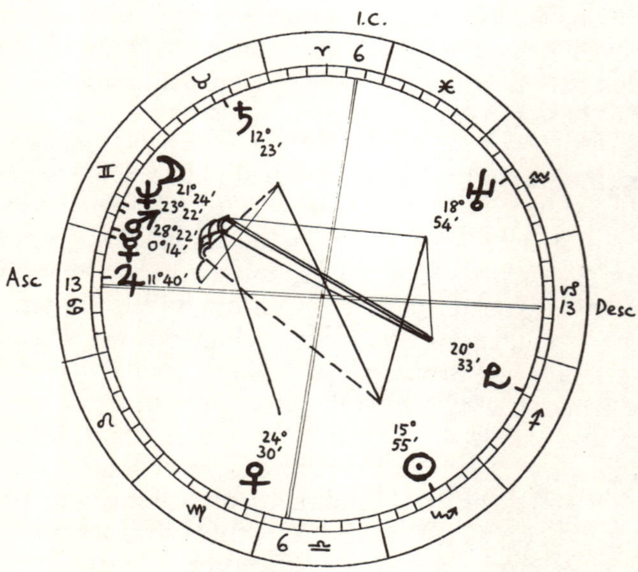

Richard Nixon: H7 chart

and structures. But, for such people, the H7 chart is a clue to the unconscious motives that underlie a person's behaviour.

Thus we can already see, from Le Corbusier's case, how the H7 chart can contain highly dangerous forces. The main danger seems to come when a person regards the inspiration which comes to him from the links in his H7 chart as being a vision of reality. If he keeps the vision to himself, but lives out his life according to this distorted picture of reality, he will probably seem to other people to be 'living in a world of his own', disconnected from the real world. But if he has the power to impose his vision on the real world through his behaviour, turning his fantasy into true reality, then the results can be quite alarming.

Politicians

There are many examples of this, especially in the charts of politicians. Perhaps the most famous is **Adolf Hitler,** whose H7 chart (p.74) has already been discussed by Charles Harvey:

> In this chart we seem to see at one and the same time Adolf Hitler's inspiration of an expanded German Empire attained by the expansion of military might, together with the iron discipline of a man who was almost inhuman in his capacity for concentrated self-control and ruthless purpose.[11]

Another example is **Richard Nixon,** the ex-President of the United States. Nixon's H7 chart is very much stronger than any of his other harmonic charts (p.74). The roots of his all-consuming drive towards personal power are contained in the close opposition in his radical chart between Mercury, Mars and Jupiter in the fourth house and Pluto in the tenth house. In the H7 chart the Mercury-Mars-Pluto opposition is still visible (though slightly beyond the orbs allowed), but now it is near to the Ascendant-Descendant axis, and Mercury and Mars are conjunct to Moon and Neptune which are closely opposite to Pluto. Pluto near to the Descendant seems to have become, in Nixon's fantasy, the other people who are 'out to get him', and Moon-Neptune-Mars-Mercury shows Nixon (again in his fantasy) responding to their hostility by concocting rational plans that will ensure their downfall. Also there is Sun opposite Saturn, seeming to show that Nixon was inspired by the idea of his own suffering. Thus in Nixon's H7 chart we can see the roots of the

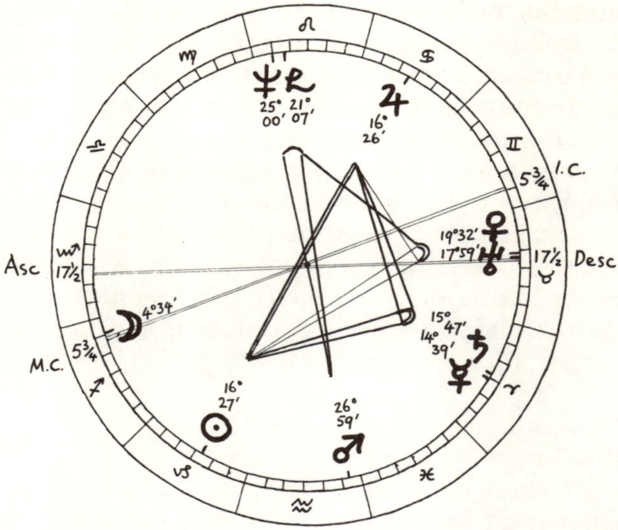

Charles Whitman: H7 chart

paranoia which led to the Watergate scandal and to Nixon's own downfall.

Nixon (whose birth time is known with great accuracy) does not, like Hitler, have a concentration of planets on the M.C. – I.C. axis. This axis in the H7 chart seems to relate to the *inspirational goals* which a person is trying to reach, whereas the Ascendant-Descendant axis relates to a person's fantasies about the present, and especially about present relationships with other people. Nixon, so far as we know, did not have any grand plans for the future similar to Hitler's; his fantasies were more concerned with his own place in the world at the present time.

Murderers

Finally, murderers and other violent criminals often have strong H7 charts, which can provide clues about their fantasy lives and hence about the motivation for their crimes. An extreme example of this is **Charles Whitman,** who went berserk and killed fifteen people, including his wife and his mother, on a single day. Whitman has an exceptionally strong H7 chart (see p.00). The dominant T-square of Sun, Jupiter, Mercury and Saturn suggests that Whitman (in his fantasy life) is reaching out towards personal glory through the effort to control and direct his mind. But the involvement of these planets with Venus and Uranus on the Descendant suggests that, in striving towards these aims, Whitman is very dependent on women, who bring adventure and romance into his life and who seem to be free of the Saturnian shackles by which he himself is bound. This is a pattern which is likely to lead to great emotional stress and to a complex love-hate relationship with women (including, presumably, his wife and his mother), but should not in itself lead to murder. But cutting across this configuration is the pattern of Mars opposite Neptune-Pluto. This pattern is far more suggestive of murder, and it seems likely that it was this pattern which was 'triggered' at the time when Whitman 'went out of his mind'.

Conclusion

Drawing from these and other cases, we can now make suggestions for the interpretation of aspects in the H7 chart.

Conjunctions
When two planets are conjunct in the H7 chart, the person is inspired by the idea of bringing them together. Although he may be unable to bring them together in his everyday behaviour, the idea of doing so has for him great emotional significance: it 'moves' him, or 'turns him on'. He may keep these feelings to himself (perhaps partly burying them in the subconscious mind), but they may also be manifested in his behaviour in various ways. He will tend to fantasize about the relationship between these planets (or rather between the forces that they represent), and he may create works of art describing these fantasies, so that the union of the planetary principles is the 'message' of the art. In his imagination he may create a romantic vision of a world in which these planetary forces are united, and this vision may become an ideal towards which he strives. Because he is inspired, he may seek to inspire other people. In some cases (perhaps as a result of strength in the other harmonics) he may be able to transform his vision into reality through his actions. The results of this can be very striking, and can often seem 'evil' because they are not rooted in reality: an unreal vision is being imposed on the real world.

Oppositions

When two planets are opposite in the H7 chart, the person is *seeking* inspiration in the idea of bringing them together. The interpretation is the same as above, except that there is more of a struggle, a determination to overcome a lack, or a restless search for inspiration. Sometimes it is clear that the person is inspired not by the idea of a union between the planets, but by the idea of a struggle between them: he may see them as representing the forces of good and evil, or light and darkness.

Squares

Here there is a still greater struggle towards inspiration, which may result in great creative achievement; but it is likely that the resulting works of art will show evidence of great emotional conflict and turbulence. Whereas in H5 there was a danger of obsessive behaviour, in H7 there is a danger of manic behaviour caused by sudden upsurges of romantic despair.

Semi-squares

These may be insignificant in themselves, unless they are linked in with two or more planets which are opposite or square to each other. When they are so linked, the planet which is semi-square to the other planets may represent a quality which the person feels is missing from the imaginative vision which he is trying to develop, and which he will constantly try to integrate into that vision.

Trines

When two planets are trine to each other in the H7 chart, the person obtains imaginative pleasure and enjoyment from the idea of bringing them together. Thus he may be transported into an imaginative world whose reality is purely subjective. It seems likely that these trines lead usually to inner daydreaming rather than to manifestation in creative action, except in cases where they are linked to a pattern of oppositions and other 'hard' aspects. In these cases the planet which is trine (or sextile) to the other planets may represent the resolution of the conflict.

Sextiles

When two planets are sextile to each other in the H7 chart,

the person is *seeking* the type of pleasurable inspiration which is associated with the trine.

Semi-sextiles and quincunxes
These are probably insignificant in themselves, except when they are features of a larger pattern and show how the person can link together the 'difficult' and 'easy' features of his H7 chart into a single unified pattern of inspiration.

In general, the seventh harmonic represents the dynamic force of man's inspiration. Whereas the fifth harmonic represents man's ability to impose on the world (and on himself) an order and structure of his own making, the seventh is concerned with his ability to impose on the world (and on himself) a meaning which has emotional significance for himself. For many people, the full potential of the H7 pattern of planetary forces is never fully realized: it may lead to daydreams and fantasies, and it may subconsciously affect the person's behaviour, but its creative potential is never explored or exploited. But when the full power of the seventh harmonic is unleashed, it can lead to savage destructiveness or to immense creative achievement. Together with the fifth harmonic, the seventh represents man's creativity: his ability to invent his own version of Reality.

CHAPTER 7
THE NINTH-HARMONIC CHART

Nine is not a prime number, but is 3 × 3. Just as the H4 chart is a map of the person's pure Twoness, so the H9 chart is a map of the person's pure Threeness. Astrologers have always understood that the number Three is to do with such qualities as pleasure, enjoyment, harmony; and so the person's capacity for developing these qualities can be seen in the H9 chart.

But it must be understood that the number Nine represents a purer or more refined type of Threeness than that which astrologers usually observe. A person's capacity for pleasure and enjoyment is perhaps shown in the third-harmonic chart, in which the trines of the radical chart appear as conjunctions. But the ninth harmonic represents 3 × 3, or 'pleasure in pleasure'.

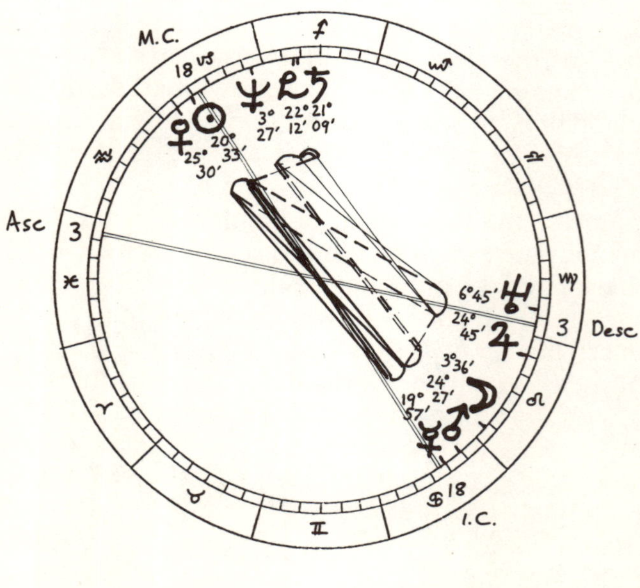

Maurice Escande: H9 chart

It represents a type of pleasure which is more profound, more difficult to experience fully, but which brings greater fulfilment. We can think of many ways to describe this. For me, the most appropriate words are *joy, happiness* and *peace*. Whereas planets linked together in the H4 chart are at war with each other, planets linked in the H9 chart are at peace with each other, because they bring each other joy and happiness.

The H9 chart, then, shows a person's capacity for joy and happiness, which arises from being at peace with himself and the world. But also (because the two things are inseparable) it shows the person's capacity to spread joy and happiness to those around him.

An example of this is **Maurice Escande,** the director of the Comédie Francaise. Gauquelin says of him:

> He ran the famous theatre with great tact, courtesy, diplomacy and effectiveness. Escande was suffering from cancer towards the end of his life, but he never paraded his pain and 'he practised heroism with little fuss' . . . . Escande won the admiration and love of all his fellow actors, and it is reported that when Elizabeth II of England met him she said 'I have just encountered that most rare of beings. A happy man.' Escande had a talent for being happy and for spreading happiness. It was perhaps his finest legacy.[1]

As we would expect, Escande has an extremely strong H9 chart (see p.80), with all ten planets linked together in a single pattern. But the close opposition between Sun-Venus and Mercury-Mars on the M.C.-I.C. axis shows that Escande did not attain happiness without a struggle. Happiness was a goal which he fought towards, rather than a spontaneous state. Nevertheless, his capacity for it was greater than average.

Happiness, and the pursuit of happiness, is the simple or secular interpretation of the ninth harmonic. But there is plenty of evidence that the ninth harmonic can also be given a more profound interpretation. For what is it that brings happiness? In order to be happy and joyful, a man must be at peace both with himself and with the world. In the fourth harmonic he was at war with the world; in the fifth harmonic he was trying to impose on it his own order and structure; and in the seventh harmonic he was trying to impose on it his own imaginative meaning. But in the ninth harmonic he *accepts* the world for what it is. He *knows* what it is, and does not try to make it different. And yet he also accepts the limitations of his

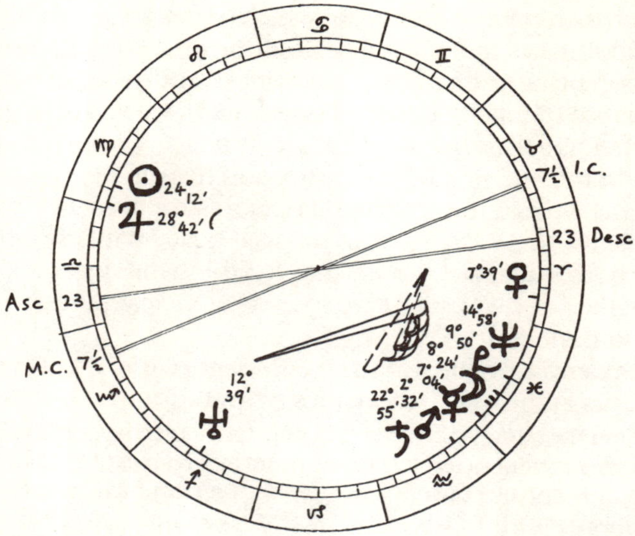

Rudolf Steiner: H9 chart

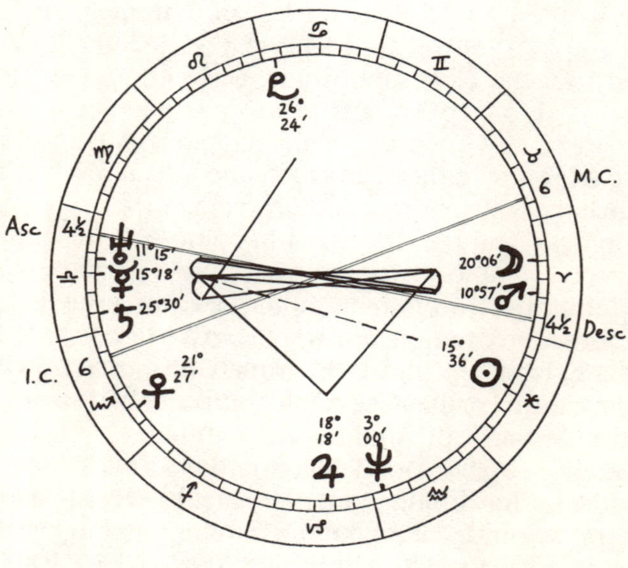

Edgar Cayce: H9 chart

knowledge. In the deepest sense, the ninth harmonic is to do with *knowledge* and *understanding* and *truth*.

And also, in order to be at peace with himself, he must accept his own place in the world, no matter how insignificant. The arrogance and self-assertion of the previous harmonics are gone, and their place is taken by humility. Nine is not the number of man, but the number of the cosmos. Within it, man has to take his place within the cosmic scheme.

Psychic powers

But knowledge also brings power. If a man has a true understanding of the world, he can exert power within it. Each of the harmonics has its own type of power; but Nine is perhaps the most truly powerful of the numbers that we have considered, even though its power is essentially non-assertive.

It is certainly remarkable how religious leaders and visionaries, and people with clairvoyant, healing, or other types of psychic power, tend to have very strong H9 charts. As an example, we can take **Rudolf Steiner,** the great spiritual leader and founder of Anthroposophy, whose H9 chart (p.82) has already been discussed by Suzanne Lilley-Harvey.[2] Steiner's statement about knowledge (quoted at the end of Lilley-Harvey's article) is an excellent statement of the true knowledge which is the province of the ninth harmonic:

> Knowledge pursued merely for the enrichment of personal learning and the accumulation of personal treasure, leads you away from the path: but all knowledge pursued for growth to ripeness within the process of human ennoblement and cosmic development, brings you a step forward.[3]

In fact, Rudolf Steiner (like Goethe, whom he admired greatly) is a remarkable example of someone who is strong in *all* the harmonic charts. But the ninth harmonic is unquestionably the most striking. It contains a sextuple conjunction of Saturn, Mars, Mercury, Moon, Pluto and Neptune, grouped around a very close conjunction of Moon, Mercury and Pluto. This is one of the most massive conjunctions that I have seen in any harmonic chart, and we can deduce that it was through the coming together of these six planets in the harmonic of knowledge that Steiner acquired his clairvoyant powers and his insights into cosmic and human truth. The truths which Steiner revealed are essentially

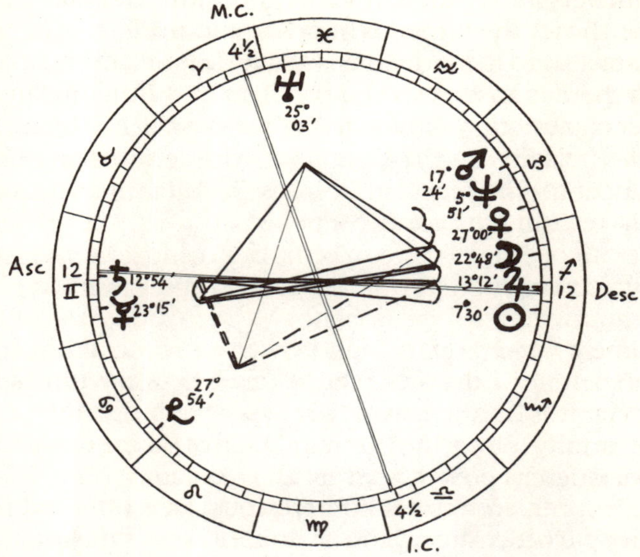

Uri Geller: H9 chart

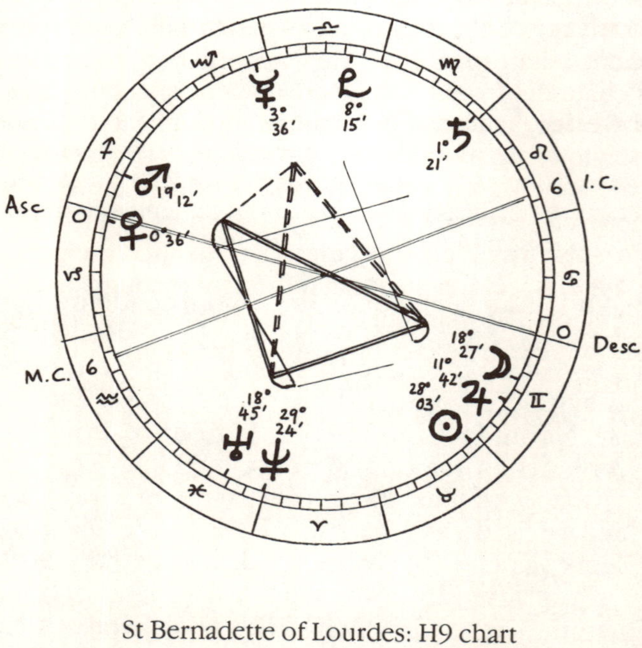

St Bernadette of Lourdes: H9 chart

of a universal (Neptunian) variety; but the Uranus-Neptune square shows the efforts which Steiner made to make them relevant also to the individual human personality.

Yet there is an unresolved problem in the chart. The Sun-Jupiter conjunction shows Steiner's acceptance of himself as a spiritual teacher and leader, and I believe that it was this conjunction which enabled Steiner to reveal to the world truths which (he claimed) had been known before to initiates but had previously been kept secret. Yet the lack of integration between this Sun-conjunction and the massive Moon-conjunction reveals a dilemma which, it seems, has not yet been resolved within the movement which Steiner founded. Do we revere the man who revealed these truths to us? Or do we revere the truth itself, of which he was only a messenger?

We can also look at the H9 chart (p.82) of **Edgar Cayce,** who has been described by Oken[4] as 'the most important Western seer, prophet and healer to appear so far in the twentieth century'. The cluster of five planets near the Ascendant-Descendant axis (Mercury-Saturn-Uranus opposite Moon-Mars, with Uranus exactly opposite Mars) shows the enormous strength of the H9 power which Cayce was able to direct out on to the world, so as to effect healing. But the non-involvement of the Sun (except through the quincunx with Mercury, which is probably not of great significance) shows that Cayce's personal fulfilment was not involved. He was content to regard himself as the channel through which cosmic energy could flow.

Uri Geller, who can bend metal through psychic power, shows a ninth-harmonic pattern which is very similar to Cayce's (with the same Mercury-Saturn conjunction on the Ascendant, and Moon close to the Descendant – see p.84). But in this case the Sun is also involved, showing that Geller, to a greater extent than Cayce, is seeking personal fulfilment through his psychic achievements. The particular nature of Geller's achievement is well shown by the exact opposition between Saturn on the Ascendant and Jupiter on the Descendant, and also the close Mercury-Moon opposition: by controlling and directing his own psychic powers through the power of thought, Geller can cause external objects to be flexible and free themselves from restraint.

Nevertheless, the close similarity between Geller's and Cayce's H9 charts leaves little doubt that Geller could, if he wished, have been successful as a healer, and Cayce would have been successful in bending metal. The differences between them would have to be sought in their radical charts and their other harmonic charts.

Another remarkable case is that of **Saint Bernadette of Lourdes**. Her H9 chart (p.84) contains a very close T-square of Moon (conjunct Jupiter), Mars and Uranus, which seems to show her openness to sudden and striking revelations of a psychic or supernatural nature; and the equally close semi-squares to Mercury show her determination to communicate these experiences to others.

In all these cases of people with psychic powers, there is a strong Moon-emphasis in the H9 chart. So it would seem that it is especially the Moon in the H9 chart which puts a person in touch with hidden forces of which other people are unaware.

A slightly different case is that of **Satya Sai Baba,** the Indian miracle-worker. Sun and Venus are on the Ascendant (with Saturn) in Sai Baba's radical chart, but in the H9 chart they are also conjunct Jupiter, trine Neptune, and sextile Mercury, and linked in a pattern with all of the other planets. In this case it seems to be the sheer joyous self-confidence of Sun, Venus and Jupiter on the H9 Ascendant which enables Sai Baba to realize his divine nature. However, the Moon is still strong through the semi-squares with Neptune and Mercury and the trine to Uranus.

Thus the H9 chart may provide an indication of whether a person is able to develop psychic powers. But, for people who

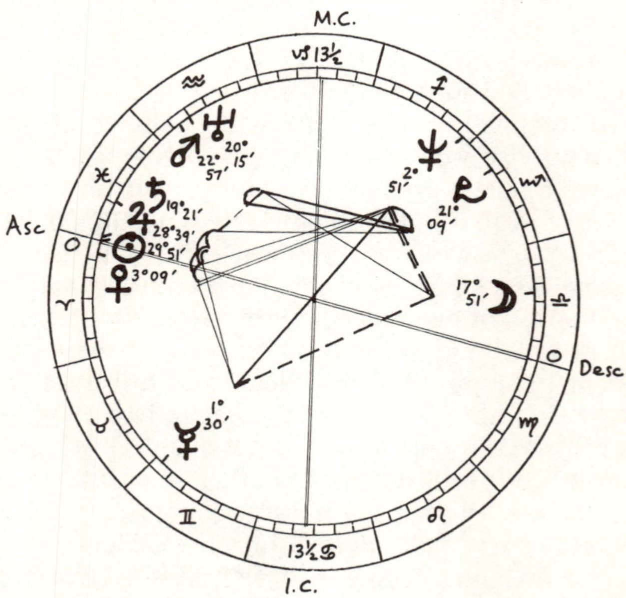

Satya Sai Baba: H9 chart

lack (or feel that they lack) these powers, the interpretation of the H9 chart can be difficult. This is especially true if one studies the charts of famous people, since most famous people are known for their active achievements and not for their states of passive enjoyment. The ninth harmonic is related, as we have said, to a person's capacity for joy and happiness; but joy and happiness are elusive things, and the goal which a person is actively pursuing is probably more likely to be a fourth-harmonic goal of victory, a fifth-harmonic goal of order, or a seventh-harmonic goal of inspiration, rather than a ninth-harmonic goal of joy.

Thus I cannot wholly accept John Addey's suggestion that the H9 chart represents 'the ideal to be realized' and that it 'stands in relation to the radical map "as the fruit to the tree" '.[5] The ideal or purpose towards which a person is striving, and the product of his life's work, are often more clearly seen in one of the other harmonic charts.

Nevertheless, the H9 chart does represent a type of ideal. It represents the highest goal towards which a person *can* strive, the goal which will bring him the greatest joy and peace and understanding; but, in order to reach this goal, it seems that he must cease the restless striving which is associated with the other harmonics, and 'open himself up' to the world in a more passive way. In this sense, the H9 chart would seem to be very important for astrological counselling.

Again, I cannot wholly accept the traditional Indian view that the H9 (Navamsa) chart represents the marriage partner.[6] Yet the ninth harmonic is clearly related not only to one's personal capacity for joy, but also to one's capacity to spread joy to those around one. Therefore the H9 chart may well provide clues about the type of person to whom one can bring the greatest happiness, and who can give one the greatest happiness in return, and so it may say something about the *ideal* marriage partner.

In the remainder of this chapter we will look at a few more examples, in the hope of further clarifying the meaning of the H9 chart.

The pursuit of enlightenment

Dr Albert Schweitzer has a strong H9 chart (p.88), but it is made up of strength which was already apparent in his radical chart. In the radical chart there is a close Sun-Mercury conjunction and also a very close T-square of Saturn opposite

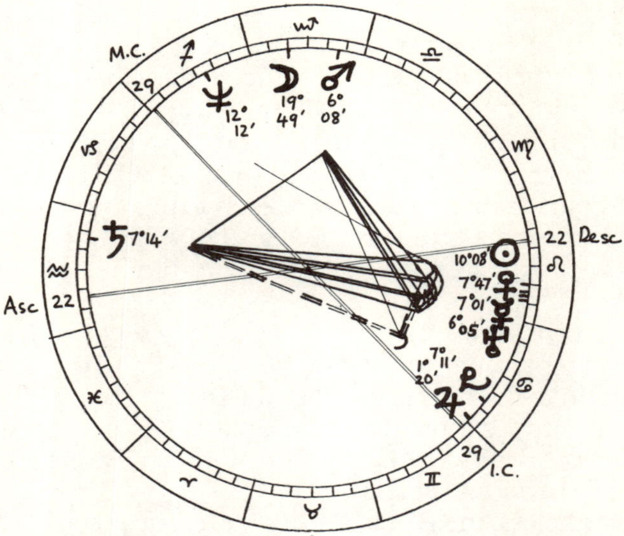

Albert Schweitzer: H9 chart

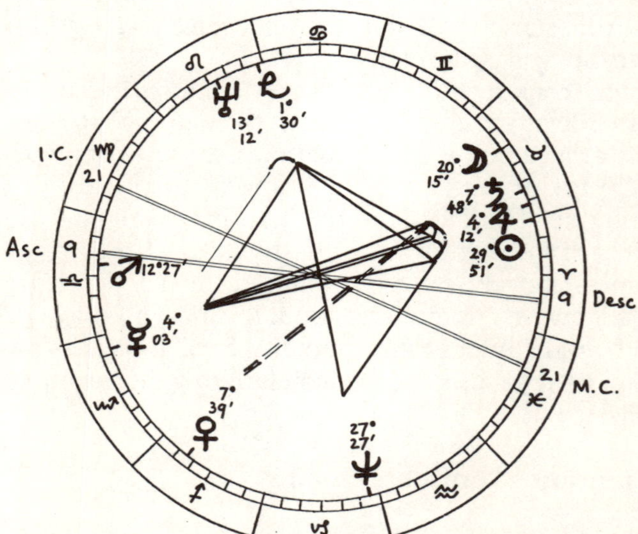

Carl Gustav Jung: H9 chart

Uranus and square to Mars. But in the H9 chart these two configurations come together with each other and also with Venus, so that Sun (near to the Descendant) is conjunct Mercury, Venus and Uranus, opposite Saturn, and square to Mars. This shows that Schweitzer can attain joy and fulfilment by acting out the forces of the radical T-square in a ninth-harmonic fashion: that is, by devoting himself to spiritual development and the service of mankind.

Carl Gustav Jung also has a strong H9 chart (p.88). Sun is closely square Neptune in the radical chart, but in the H9 chart Sun and Neptune come together with Mercury, Jupiter, Saturn and Pluto to form a Grand Cross. The involvement of the Sun with four of the five outer planets shows Jung's ability to develop a deep and universal understanding of the human condition; and the involvement of Mercury shows his ability to express this understanding in words and communicate it to others.

In Jung's case, as in Schweitzer's, the emphasis is on the Sun rather than the Moon. The Moon in H9 is, as we have said, related to psychic powers; the Sun does not carry this connotation, but is concerned rather with the development of personal enlightenment and understanding. Sun conjunct Jupiter (which we saw also in the case of Rudolf Steiner) shows Jung's acceptance of himself as a teacher or enlightener of others; but, since Sun is also conjunct Saturn, Jung also feels that he must pull his teachings down to earth, and be concerned with the mundane details of human life and suffering, as well as with transpersonal reality.

Also we can see that, in Jung's case as in Schweitzer's, the emphasis is on oppositions and squares. These 'hard' aspects introduce an element of Twoness into the pure Threeness of the ninth harmonic, and show that there is a struggle to attain enlightenment. (For Rudolf Steiner, on the other hand, with his massive conjunction in H9, spiritual development came easily: Steiner's struggles were in the other harmonics, not in the ninth.)

Creative artists

We can also look at the ninth harmonic in the charts of creative artists. In the previous chapter we said that the fifth and seventh harmonics represent creativity in the sense of inventiveness: the ability to invent new versions of reality. But the ninth harmonic also plays a role in creativity. It represents the ability to submerge oneself in one's surroundings and to obtain from them the H9

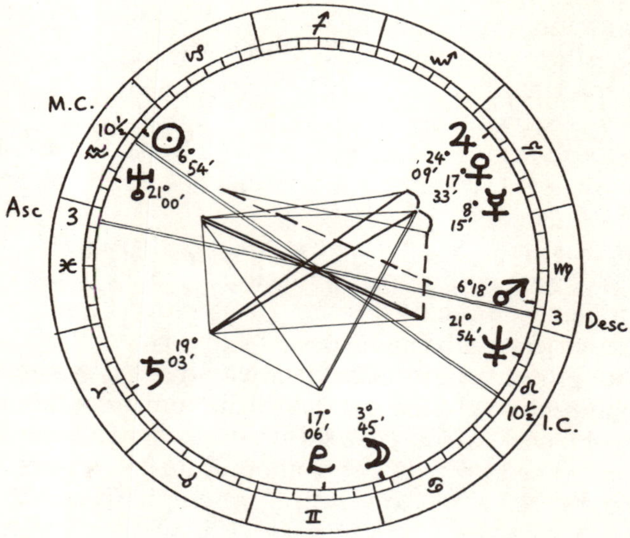

Camille Corot: H9 chart

qualities of joy and peace and understanding, so that one can then pass these qualities on to one's fellow-men through a process of *re*-creation. An artist whose main strength is in the ninth harmonic is therefore acting as a *channel* for forces which he finds outside himself, rather than creating reality afresh in his own image.

Also, since the ninth harmonic is concerned with bringing happiness into other people's lives, an artist who is working out of the ninth harmonic may well seem to be lighter in tone, more concerned with pleasure and even with entertainment, than one who is working out of the fifth and seventh harmonics. An artist who is trying to impress the world with the grandeur of his vision and the originality of his inspiration or his style is more likely to derive his strength from the fifth and seventh harmonics. In practice, however, there are no completely clear-cut cases of fifth-, seventh-, or ninth-harmonic persons: everybody has some degree of strength in all of the harmonics, and their effects are constantly interacting. Therefore, in order to carry out a full assessment of an artist (or of any other person) we have to look at all the harmonics together.

One example of a painter whose greatest strength is in the ninth harmonic is the landscape painter **Camille Corot**. Corot's

H9 chart (p.90) contains a Grand Trine of Venus, Jupiter, Uranus and Pluto, with Venus-Jupiter also opposite Saturn, and Uranus also opposite Neptune; also Sun is trine to Mercury which is conjunct Venus (so that Venus is the pivotal planet of the chart). Corot's struggle throughout his life was to re-create in paint the beauty of what he saw before him:

> The quality of his painting was due to this direct contact with nature, his observation of light, his judgment of colour-tone values, his sense of structure and his feeling of atmosphere.[7]

> He did not let realism stand in the way of an elusive poetry and beauty: through the truth, he still pursued something intangible. Before dying he confessed: 'It seems to me that I never learned how to paint sky. What is there before me is much more beautiful, much deeper, much more transparent'.[8]

This awareness of inadequacy in the effort to translate heavenly beauty into mundane paint is well conveyed by Venus-Jupiter opposite Saturn in the H9 chart.

Charles Dickens is an example of a writer whose greatest strength is in the ninth harmonic. His H9 chart (p.92) contains a complex pattern of planets, dominated by trines and sextiles, but with an exact Venus-Jupiter opposition. Dickens, like Corot, tried to record exactly what he saw, and, more directly than Corot, he was motivated by a desire to bring happiness into the lives of other people. A contemporary critic wrote:

> No one thinks first of Mr Dickens as a writer. He is at once, through his books, a friend. He belongs among the intimates of every pleasant-tempered and large-hearted person . . . . It is not in his purely literary character that he has done most for us, it is as a man of the largest humanity, who has simply used literature as the means by which to bring himself into relation with his fellow-men.[9]

And Dickens himself said:

> I have great faith in the poor; to the best of my ability I always endeavour to present them in a favourable light to the rich; and I shall never cease, I hope, until I die, to advocate their being made as happy and as wise as the circumstances of their condition, in its utmost improvement, will admit of their becoming.[10]

This desire to relieve the suffering of others is well shown by

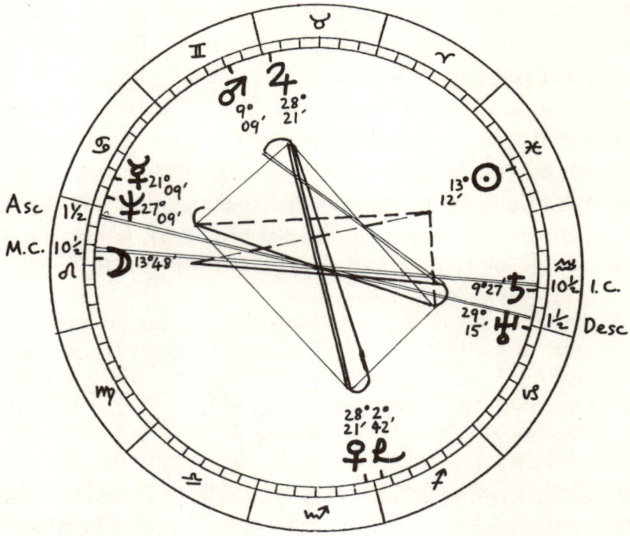

Charles Dickens: H9 chart

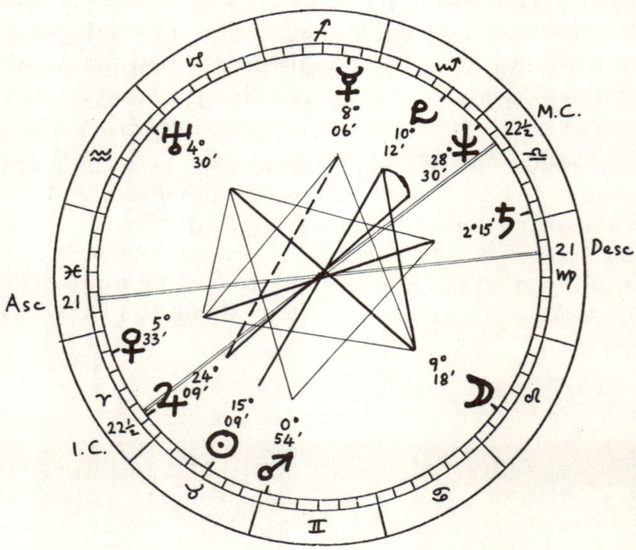

Peter Sellers: H9 chart

Moon opposite Saturn, close to both the Ascendant-Descendant and M.C.-I.C. axes, in Dickens's H9 chart.

A confused pattern

Lastly we will look at quite a different kind of case: the H9 chart of the actor **Peter Sellers** (p.92). There are a large number of links in this chart: in fact there are two separate Grand Trines (Moon-Mercury-Venus and Mars-Saturn-Uranus) which clearly gave Sellers the capacity for a great deal of joy. But the Grand Trines are facing in opposite directions, like a Star of David; and criss-crossing them are four separate oppositions (Sun opposite Pluto, Moon opposite Uranus, Venus opposite Saturn, and Jupiter opposite Neptune). In fact, if wer were to allow slightly wider orbs, Mercury would also be opposite Mars, and we woild then have all then planets involved in five separate oppositions.

Thus the overall impression conveyed by this chart is one of confusion and lack of direction: and this is also the impression conveyed by Sellers's biography. Peter Sellers was not weak in the ninth harmonic: he knew what happiness was, but he did not know where he could find it: 'From an early age religion fascinated Peter Sellers, puzzled him, and pursued him through the years'.[11] He tried the Church, he tried spiritualism (it seems probable that he possessed some psychic powers), but he did not try anything for long. Mostly he looked for happiness through becoming other people, and his great receptiveness to the qualities of other people is perhaps conveyed by the Grand Trine of Moon, Mercury and Venus.

Conclusion

Using these and other cases, we can now make suggestions for the interpretation of aspects in the H9 chart.

Conjunctions

When two planets are conjunct in the H9 chart, the person is able to experience joy, peace and happiness in the relationship between these two planetary principles, as they manifest both in his own personality and in the outside world. In so far as he experiences joy and happiness, he is able to spread it also to other people. This joy and happiness seems to result from an understanding of the true relationship between these planetary principles: he accepts this as it is, and does not try to change it. This understanding and receptivity can lead to psychic powers, at least when the Moon is involved. It can also lead to artistic activity, in which the aim is to spread happiness by re-creating the truth and beauty which is seen in the world, rather than to create reality afresh in one's own image. Depending on the nature of the planets involved, the ninth harmonic can also be connected with religious faith and spirituality. It can also have the qualities of innocence and humility.

Oppositions

When two planets are opposite to each other in the H9 chart, the person is *striving towards* the joy and peace which would result from the conjunction. The interpretation is the same as above, except that the happiness is attained through effort rather than occurring spontaneously. Sometimes one of the planets is externalized, and represents the type of happiness or fulfilment which the person hopes to impart to other people through the operation of the other planet.

Squares

Here the sense of struggle is still stronger: squares in the H9 chart show the person actively struggling towards a type of happiness, faith, or enlightenment. Sometimes they may be connected with sudden spiritual revelations.

Semi-squares

These may be insignificant in themselves, unless they are linked in with two or more planets which are opposite or square to each other. When they are so linked, the planet which is semi-square to the other planets may represent a quality which the person feels is missing from his own search for enlightenment, but which he will constantly try to integrate with the other planets so as to help him in his search.

Trines

These represent a still purer type of pleasure, which (depending on the nature of the planets involved) can be called ecstasy, bliss, or serenity. In practice, however, it is often difficult to differentiate between the interpretation of the trine and that of the conjunction.

Sextiles

When two planets are sextile to each other in the H9 chart,

the person is *striving towards* the realization of the type of pure bliss which is represented by the trine.

Semi-sextiles and quincunxes
These are probably insignificant in themselves, except when they are features of a larger pattern and show how the person can link together the 'difficult' and 'easy' features of his H9 chart into a single unified pattern of enlightenment.

For 'ordinary' people, who do not possess psychic powers or the ability to experience mystic joy (or who perhaps are not ready to be told that they possess these things), the best way to interpret the H9 chart seems to be to regard it as a clue to the *way in which a person can find happiness.* The problem with the ninth harmonic is that it seems to be recessive: some people do not realize the full potentialities of their H9 charts until shortly before their deaths, and other people (perhaps the great majority) never fully realize them. Thus we can look at the H9 chart of a person such as Richard Nixon, and see in it aspects which seem to bear no relation to his personality: it is as though he had never learnt to live in the ninth-harmonic mode. But people desire happiness none the less, even though they may not know how to find it; and, for those who are willing to learn, the H9 chart can be of tremendous value.

But it is clear that the ninth harmonic goes beyond this, perhaps into areas which are difficult to express in words. In fact, in investigating the ninth harmonic (even more than the other harmonics), I feel a strong sense of excitement and also of awe, as though I am treading on holy ground, and am circling around the edge of some of the great secrets of the universe. Certainly I do not feel that I have taken more than a few faltering steps towards the centre of the mystery. But I hope that this book may at least inspire some readers to investigate further and (in the true spirit of the ninth harmonic) to share their findings with the world.

CHAPTER 8
GENERAL HINTS ON INTERPRETATION

In the previous four chapters I gave some examples of harmonic charts, but did not analyse them in great detail. In this chapter I will make some general points about how harmonic charts should be analysed and interpreted, before going on to deal with aspects between specific pairs of planets. I will assume in this chapter that the astrologer is not interested in interpreting harmonic charts *for their own sake* (as was the case in Chapters 4-7), but is interested in *integrating* the interpretation of the harmonic charts into the total analysis of the subject's personality.

Use of the radical chart

The first, and perhaps the most important, point is that the harmonic chart should always be interpreted against the background of the radical chart (the birth chart itself). The radical chart should be interpreted first, and the harmonic chart should be regarded as presenting *additional* information which must be fitted into the picture presented in the radical chart, rather than as presenting a totally 'new' picture of the subject. The planets which are shown in the harmonic chart are still the *same* planets that were shown in the radical chart, and their true position in the heavens (by sign and house) is as shown in the radical chart. Therefore, when looking at a harmonic chart, we should always have the radical chart alongside it, so that we can constantly cross-check between the two.

In the radical chart it is particularly important to note the *close aspects* and *unaspected planets*.

Close aspects

When an aspect in the radical chart is very close to exactitude, it will recur throughout the series of harmonic charts. For example, **John McEnroe,** the tennis player, has a close Mars-

Pluto square in his radical chart, which recurs in all his harmonic charts and comes into contact with different planets in each case: in the H4 chart Jupiter is conjunct both Mars and Pluto; in the H5 chart Venus is conjunct Mars and square to Pluto; in the H7 chart Moon is semi-square both Mars and Pluto; and in the H9 chart Mercury is trine Mars and quincunx Pluto. Thus we can see that the interpretation of this one aspect is absolutely central to an understanding of McEnroe's personality. We should start by interpreting it as it appears in the radical chart: Mars (conjunct Moon) in Gemini in the eighth house, closely square to Pluto (opposite Sun/Mercury, square Jupiter, trine Saturn) in Virgo in the eleventh house. When we have understood this, we can then go on to consider the ways in which Mars-square-Pluto is linked to other planets in the harmonic charts.

Sometimes the feature which recurs in all the harmonic charts is not a single aspect, but a configuration of several aspects. For example, **Bobby Fischer,** the chess champion, has a very close Grand Trine of Mars, Uranus and Neptune in his radical chart, which recurs in all the harmonic charts and connects with different planets in each harmonic. (In the H4 chart Moon is conjunct Mars, trine Uranus and Neptune; in the H5 chart Sun is conjunct Uranus, trine Mars and Neptune; in the H7 chart Pluto is square Uranus, semi-sextile Neptune and quincunx Mars; and in the H9 chart Mercury is trine Mars, Uranus and Neptune.) Once again we must interpret the Grand Trine as it appears in the radical chart (Mars in Aquarius on the Descendant opposite Pluto, trine Uranus in Gemini in the eleventh house, trine Neptune in Libra in the third house), and then go on to consider the additional information from the harmonic charts.

Unaspected planets

When a planet is unaspected in the radical chart, it can be a very powerful force in the personality, as Geoffrey Dean has shown.[1] But, since the planet has no links with other planets in the radical chart, we have to look to the harmonic charts to find the ways in which the person integrates this planet into his personality. For example, **Eleonora Duse** (whose radical and harmonic charts we will discuss in Chapter 12) has Mars unaspected in the radical chart, and weakly aspected in the H4 chart also; but in the H5 chart Mars is square Pluto, sextile Moon, and semi-square Sun, Jupiter and Uranus. Hence, Duse's Mars has a particularly fifth-harmonic quality, since H5 is the first harmonic in which it is integrated with other planets.

This applies not only to unaspected planets but also to weakly-aspected planets (e.g. planets which in the radical chart have only one aspect which in itself is not very strong, such as an inexact sextile or quincunx), and also to unaspected duets (i.e. pairs of planets which in the radical chart are linked to each other but to no other planets). The harmonic charts can often show the ways in which these unaspected duets are linked with other planets in the chart. Sometimes we can see that two separate duets (or larger groupings) are brought together in a particular harmonic chart. For example, in Chapter 6 we saw that **Le Corbusier** has both Sun conjunct Uranus and Moon conjunct Pluto in his radical chart, but in the H7 chart these two groupings are brought together in a pattern of aspects (Sun-Uranus opposite Moon-Pluto) which is linked also with Jupiter and Saturn.

Patterns of aspects

Within the harmonic chart, it is important to look for patterns of aspects: that is, groups of three or more planets that are linked together in an aspectual pattern. These patterns (or configurations) can be of many kinds. For instance, if there are only three planets in the group, they may be linked together in a triple conjunction, or with one conjunction and two oppositions, or in a T-square, or in a Grand Trine, or with a square and two semi-squares, and so on. If there are more than three planets in the group, the number of types of patterns is still further increased.

The art of interpreting harmonic charts is, in my opinion, very much the art of identifying these configurations and interpreting them *as a whole*. It is not necessary to interpret *every* aspect in a harmonic chart: in fact, a single aspect in a harmonic chart may not be very important if it is not linked to other aspects. But a configuration which binds together a large number of aspects is an important force, showing that the person is able to *integrate* these planetary forces in the manner of this particular harmonic number. The principle of Fiveness will be very much more important for a person who has a large number of linked aspects in his H5 chart than for a person who has very few.

The nature of the integration will of course depend on the identity of the planets involved in the pattern, and it will also depend on the type of aspects which they form. If three planets are in conjunction in the H5 chart, their relationship will have the nature of pure Fiveness. But if they are in a T-square, it will be a more difficult relationship in which they are 'straining towards Fiveness', and if they are in a Grand Trine it will be a more joyful relationship in which they find 'pleasure in Fiveness'.

As an example, we can take the H9 chart of **Carl Gustav Jung,** which we discussed briefly in Chapter 7 (p.88). In this chart, the Mars-Uranus sextile can probably be ignored, since it is is not linked to other aspects; and the wide Uranus-Pluto conjunction and the Venus-Saturn quincunx (even though it is close) are also relatively unimportant. But the Grand Cross involving Sun, Jupiter, Saturn, Mercury, Neptune and Pluto is clearly extremely powerful, and enables us immediately to say that 'straining towards Nineness' (that is, towards enlightenment, peace and happiness) was a central feature of Jung's personality. If we want to understand this Grand Cross in detail, it is important that it should be understood *as a whole* rather than as a series of disconnected aspects. Since this is such a complex configuration, its interpretation may be a lengthy process, and may need to be tackled in stages. We might, for instance, start by trying to understand the triple conjunction of Sun, Jupiter and Saturn; then consider the oppositions between these planets and Mercury; and finally consider the squares to Neptune and Pluto. Throughout this process we should bear in mind the planets' sign- and house-positions in the radical chart, and we should also note which of the aspects within the Grand Cross have been carried over from the radical chart. If we were to do all this, we would have a very full understanding of the Nineness within Jung's personality.

Close aspects

The closer an aspect is to exactitude, the stronger will be its effects. Therefore, in the harmonic charts as in the radical chart, it is very important to note close aspects. (Using double lines to draw close aspects, as recommended in Chapter 3, is a major aid to their identification.) A close aspect often seems to have a compulsive effect even when it is not linked with other aspects in a configuration. For instance, **Winston Churchill's** H4 chart (p.40) contains two close squares (Sun square Uranus and Moon square Venus) which seem to have had a major effect on his personality, even though neither of them is part of a configuration.

If a close aspect is part of a configuration, it can be even more important. If we look again at **Jung's** H9 chart, we can see that the closest aspect in the Grand Cross is Mercury opposite Jupiter. This, then, is the strongest or most compulsive aspect in Jung's 'straining towards Nineness'. In Chapter 10 the following interpretation is suggested for Mercury-H9-Jupiter: 'Delighting in one's open-mindedness and freedom from restrictive thought-patterns: rejoicing in one's ability to spread good cheer through communication'. Since Mercury is *opposite* Jupiter in Jung's H9 chart, Jung is *striving towards* this type of delight. This interpretation (which may need to be modified in the light of the sign- and house-positions of Mercury and Jupiter in Jung's radical chart: Mercury in Cancer in the sixth house; Jupiter in Libra in the eighth house) then becomes the 'centrepiece' of the analysis of the Grand Cross in Jung's H9 chart.

Occasionally we may find a configuration which consists entirely of close aspects. Examples of this are the T-square of close aspects between Sun, Mercury and Venus in the H5 chart of **Barry Humphries** (p.56), and the T-square of close aspects between Moon, Mars and Uranus (linked also to close semi-squares to Mercury) in the H9 chart of **St Bernadette of Lourdes** (p.84). Such configurations are of course extremely powerful, and usually their effects are clearly and strikingly visible in the subject's behaviour. When a person has a 'close configuration' in one of his harmonic charts, his life often seems to be dominated by the type of behaviour associated with that harmonic.

Pivotal planets

Sometimes we find that a harmonic chart contains two or more configurations which are linked together by a single planet which forms part of all of them. This planet can be regarded as the pivotal planet of the chart, and often seems to act as a dominant force within the type of behaviour which is represented by that harmonic chart. We have seen several examples of this in Chapters 4-7; one example is Mars in the H7 chart of **Salvador Dali** (p.68), which links together the Mars-Moon-Mercury conjunction, the Mars-Jupiter-Neptune T-square, and the Mars-Saturn-Uranus sextiles. Mars in H7 appears to be connected with erotic fantasy, and this is certainly a dominant feature of Dali's imaginative world.

Planets close to the Angles

If the birth time is known to be reasonably accurate, it is worth noting any planets which are close to the Angles of the harmonic chart, since these appear to play a more prominent and visible part in the person's response to the harmonic number than those which are not. In particular, an opposition between planets on the Ascendant-Descendant axis or the M.C.-I.C. axis (such as, for instance, the opposition between Sun-Mercury-Mars and Jupiter-Uranus on the Ascendant-Descendant axis of **Mozart's** H5 chart, p.52) often seems to be particularly strong. The interpretation of planets on the Angles of harmonic charts will be discussed in Chapter 11.

Interpretation of aspects between planets

Chapter 10 offers interpretations for aspects between each pair of planets in each of the four harmonics: H4, H5, H7 and H9. However, these should only be regarded as starting-points for an interpretation. The precise way in which a particular aspect is interpreted in a particular harmonic chart will depend (as we have already seen) firstly on the other planets with which it is joined in a pattern or configuration of aspects, and secondly on the houses and signs which the planets occupy *in the radical chart.* With the aid of this information, it is often possible to reach an interpretation of a harmonic aspect which is very precise and very specific to the individual case.

To illustrate this, let us take one particular harmonic aspect – Mercury conjunct Saturn in H7 – and see how the interpretation of it will vary, depending on the other planets involved and on the houses and signs occupied in the radical chart. The general interpretation given for this aspect in Chapter 10 is: 'Inspired by the idea of controlling and disciplining the mind, and of thinking and communicating in a down-to-earth and practical way: inspired by the idea of scientific proof and experimentation, and of belief systems that limit one's freedom of thought: inspired to communicate one's limitations and obligations, and those of other people'.

To make this general interpretation more concrete, we can take some examples of famous people who have a close Mercury-Saturn conjunction in H7. Firstly, **John Addey** had this aspect in his own chart, and suggested that it was connected with an interest in the mystical nature of Number.[2] Mercury in Addey's radical chart is in Cancer in the twelfth house in conjunction with Pluto, which suggests a mind which probes deeply and intuitively into the secrets of the universe; and Saturn is in Virgo in the second house, suggesting that Addey satisfied his need for control and limitation by analytical methods and by investigating his surroundings. The inspirational link between Mercury and Saturn gives Addey an interest in the ultimate laws which control and limit the behaviour of people and things on Earth.

Secondly, the impersonator **Mike Yarwood** also has Mercury closely conjunct Saturn in H7. Mercury is again in Cancer, but in Yarwood's case it is conjunct Venus in the first house, so that it is more concerned with charming, seductive and confident communication with other people in order to win their love and admiration. But Saturn is in Taurus in the twelfth house in conjunction with Uranus, and this is a placement which makes Yarwood acutely aware of his personal limitations and his difficulties in attaining true individuality. The inspirational link between Mercury and Saturn helps Yarwood to link together these two sides of his personality: he communicates confidently, but he also communicates his sense of personal inadequacy and his preference for being in other people's shoes: and (helped by other factors in his chart) he disciplines his mind to be concerned not with the communication of self, but with the communication of other.

Thirdly, the fake-artist **Tom Keating** also has a close Mercury-Saturn conjunction in H7. Keating is a strange case of a painter who has brilliant and original gifts but who seems to be compelled always to copy the styles of other painters and to have the results accepted as genuine Old Masters. Keating's original and inventive mind is shown by Mercury in Aquarius in close conjunction with Uranus (and wider conjunction with Venus): Keating (like Yarwood) wants to charm and seduce, and he wants to do so in original and inventive ways that demonstrate his own uniqueness. But these planets are in the fourth house, and fourth-house planets are unlikely to come out into the world on their own: they need the stimulus of planets in more outgoing houses. Saturn (in Cancer) is on the Midheaven in conjunction with Neptune, showing that the goal towards which Keating is striving is the abnegation of self (this being, as we shall see in Chapter 9, the common factor between Saturn and Neptune): and the inspirational link between Mercury and Saturn shows that Keating is inspired to use his mental gifts in ways which involve the denial of self, by imposing on himself rules which prevent the expression of his individuality.

Our fourth case of a person with a close Mercury-Saturn conjunction in H7 is the writer **Herman Melville,** author of *Moby Dick*. In Melville's radical chart, Mercury is in Virgo in the fifth house and Saturn is in Aries on the cusp of the twelfth house, which is a pattern not very different from Mike Yarwood's. But in Melville's H7 chart, Mercury is in conjunction not only with Saturn but also with the Moon and the Ascendant, and opposite to Sun on the Descendant: the Moon increases the emotional element (being moved to tears by suffering), the Sun increases Melville's personal involvement at the deepest level, and the Ascendant-Descendant axis increases Melville's need to demonstrate his inspirational (H7) qualities to the world. Thus Melville is more totally caught up in the Mercury-Saturn septile than any of the other cases we have considered, and he uses its inspiration to communicate a world-view which is deeply Saturnian, concerned with duty, obligation, sorrow and suffering.

Other examples of people with Mercury-Saturn conjunctions in H7 (though not always so close as in the cases we have considered) include the comedian Kenneth Williams, the composer Giacomo Puccini, the murderer **Charles Whitman** (see p.76), the scientist Johannes Kepler, the painter Paul Gauguin, the actor Jean-Louis Barrault, the 'Playboy King' Hugh Hefner, the writers John Ruskin and Bertrand Russell, the politician Archbishop Makarios, and the spiritual leader Jiddu Krishnamurti. In each case the precise interpretation will depend on the other planets involved and on the houses and signs occupied in the radical chart.

Thus we can see that the general interpretations of aspects, which are given in Chapter 10, do not take us very far. In order to understand an aspect in a harmonic chart, we must first obtain a full understanding of the radical chart and then look at the harmonic chart as a whole and interpret it as an integrated pattern of aspects. The general interpretations can only act as starting-points.

The importance of the Sun

While investigating harmonic aspects, I have continually been impressed by the importance of aspects involving the Sun. They seem to have a far more obvious significance, and a far more all-pervading effect, than other aspects which do not involve the Sun.

If we make a list of people who share a particular aspect (as we have just done for Mercury septile Saturn), there often seems at first glance to be little or nothing in common between the people in it. This is not really surprising, since Mercury septile Saturn is only one of a large number of aspects in each of their charts, and we could not expect it on its own to 'stamp' a person as being of a particular type.

But a list of people who share a particular aspect *involving the Sun* (such as, for instance, *Sun* septile Saturn), will often clearly show a particular common quality, and an understanding of this quality will be central to an understanding of their personality. In my view (which I will expand further in Chapter 9) this is because the Sun represents the self. If a person has Mars conjunct the Sun in H7, this means that Mars is related *to the person himself* in an inspirational way, and that this colours the whole nature of his Mars, and, indeed, of his whole personality.

For this reason, even if we do not have the time or the inclination to do a full analysis of a series of harmonic charts, it may still be worthwhile to look at the *Sun's* harmonic aspects, since this will uncover some central facts about the subject's personality and will make possible a fuller and truer interpretation of the radical chart.

Synthesis of harmonic charts

These hints in interpretation are intended primarily for astrologers who are hoping to incorporate the use of harmonic charts into their counselling work. The case-studies in Part Three of the book, and especially the studies of 'Linda' and 'Susan' in Chapters 15 and 16, may provide further help. However, I will not attempt in this book to lay down a precise format for the use of harmonic charts in consultancy, since it is up to each astrologer to use them in his own way and to incorporate them into his existing methods.

One approach is to use the harmonic charts only when they contain particularly important or striking information. A quick glance at the pattern of aspects in a harmonic chart will usually reveal whether the chart contains vital information which *must* be included if the analysis is to have any semblance of completeness. Harmonic charts which contain few links between planets, or a confused and disintegrated pattern of linkages, can more safely be ignored than those which contain a strong and integrated pattern of aspects. A second approach (as I have just suggested) is to concentrate only on aspects involving the Sun. A third approach is to use the harmonic charts only when the astrologer is required to comment on the client's creativity (since this is probably the area in which harmonic charts are most completely indispensable).

Many astrologers, however, will want to make the fullest use of insights from harmonic charts in their counselling work. For these astrologers, the aim should be to *synthesize* the insights from the harmonic charts into the overall analysis of the radical chart, rather than to regard the harmonic charts as an appendage to the main analysis. To say to a client 'You have a fifth-harmonic chart which says such and such' would not only require a long technical explanation, but also make it seem that the H5 chart (and the type of behaviour associated with it) was completely separate from the rest of the chart. If the client has Sun conjunct Mars in the H5 chart, the astrologer should be aware of this when he first mentions the Martian side of client's personality, and should allow this information to colour his whole interpretation of Mars.

Astrological counselling is, however, not a science but a creative art, which can only be learnt through practice; and this applies also to the interpretation of harmonic charts. The astrologer must learn how to use harmonic charts in his own way, and must form his own judgement about whether they increase his astrological skill and his ability to help the client.

CHAPTER 9
THE PLANETS

This chapter may seem out of place in a specialist book on harmonic charts, since it deals with the interpretation of the planets in astrology as a whole. However, the statements on the meaning of the planets which are given below are based on my research into the aspects in harmonic charts.

My approach to the interpretation of harmonic aspects has been, as far as possible, empirical: for each individual aspect (for instance, Mars conjunct Pluto in H7) I have listed all the people known to me who had this aspect (noting also which of them had it as a *close* aspect) and have tried to reach an interpretation which made sense for all of them. By this process I have sometimes reached an interpretation which is different from what I would have expected, and which has caused me to revise my understanding of the nature of the planets, or of the nature of the aspect, or both.

Thus, I believe that I have emerged from this investigation with a clearer understanding of the essential underlying nature of each of the planetary forces. In this chapter I will briefly describe my present interpretation of the meaning of each of the planets in a birth chart.

The Sun

The Sun represents 'I myself as a pro-active being'. It is the core of self-awareness. It is the 'I' which speaks when I say 'I *am* such-and-such a person, I *have* certain thoughts and feelings and desires, I *did* certain things'. The other planets (apart from the Moon) represent only particular behavioural attributes of the self-force which is represented by the Sun. Aspects to the Sun show attitudes to self and routes by which we find self-fulfilment. Aspects which do not involve the Sun show attitudes, behaviour-patterns and capabilities which we possess, but which do not involve the self at the deepest level.

The Moon

The Moon represents 'I myself as a re-active being'. It is, in a sense, an intruder in the system (just as the physical Moon is not in itself an important member of the solar system, but derives its importance only from its nearness to the earth), since it represents an alternative centre of awareness, the realization of which may hamper (or interfere with) the realization of one's Sun-nature. The Moon shows not awareness of self, but awareness of environment: within our Moon-nature we do not act, but only re-act. Hence the Moon (and the aspects of other planets to the Moon) shows the nature of our reactions to people, things and events in our environment. It is concerned with our instinctive (unthinking) feelings towards things that are outside the self, and the instinctive way in which we respond to them.

The next three planets, Mercury, Venus and Mars, represent the three primary faculties by which we sort out the nature of our attitudes to ourselves and to the world and determine how we will behave. Each of them has both an internal and an external side: each represents both an aspect of our internal consciousness-processes and an aspect of our external behaviour.

Mercury

Mercury represents, internally, the process of thought, reason and logicality. Externally, it represents the process of communicating by means of words and symbols, and especially the communication of our rational thoughts and conclusions. Like Venus and Mars, Mercury (in its external mode) is concerned with a process of persuasion: Mercury is concerned to persuade other people that we are intellectually right, so as to win their agreement and also their respect and admiration.

Venus

Venus represents, internally, the process of valuing: the process of deciding what we like and what we do not like, or what we love and what we hate. Externally, Venus is concerned with conveying these likes and dislikes to other people through our behaviour. But also, Venus is concerned internally with *who* we are drawn to and *who* we want to be close to, and therefore it is externally concerned with the process of attracting and pleasing other people in order to win their appreciation, admiration and love. But it is also associated with the search for privacy, and with the process of repelling those people from whom we desire to be separate.

Mars

Mars represents, internally, the process of will: the process of deciding on goals and summoning the energy to go out and achieve them. Externally, it represents the actions by which we seek to achieve our goals, and this includes the process of attempting to bend other people to our will – the process of persuasion by power or force. But also, Mars seems to include the sheer expenditure of physical energy, and the process of trying to impress other people through physical display.

The five outer planets – Jupiter, Saturn, Uranus, Neptune and Pluto – represent not faculties, but *principles* which help to guide our actions. Essentially they are *moral* principles, concerned with the difference between good and bad, right and wrong. The five inner planets are essentially amoral, concerned only with 'this is what I am, this is what I perceive, and this is what I want'. But the outer planets give us moral direction, and the conflicts between them represent the moral dilemmas to which we are subject.

Jupiter

Jupiter represents the principle of growth and freedom. Expressed as a moral principle, Jupiter says: 'I ought to grow and expand as much as possible and as widely as possible: I ought not to accept my apparent limitations, but ought to break through into a state of freedom from limitations and into a state of total opportunity: I ought not to accept the limits of my understanding, but ought to seek an understanding which is higher, wider and deeper. I ought to be optimistic about the future, to refuse to be disheartened by setbacks, and to believe that anything is possible.'

Saturn

Saturn is the opposite of Jupiter: it represents the principle of control and limitation. Expressed as a moral principle, Saturn says: 'I ought to accept the limitations which surround me, the limitations of my capabilities and my environment, the limitations of my earthly destiny. I ought to refrain from having unrealistic expectations about what I can understand and what I can achieve. I ought to accept my lot, do dutifully what is required of me, and be realistic about the future.'

Uranus

Uranus represents the principle of individuality and originality. Expressed as a moral principle, Uranus says: 'I am a unique individual with a unique destiny: I therefore ought to draw the boundaries between myself and other people as sharply and precisely as possible, so as to maximize my own individuality and ensure that my own personal impact in the world is as great as possible. I must avoid being led astray by convention, by custom, by half-baked and confused ideals, or by anything that does not belong to me personally and does not arise from my own nature and my own potentialities. At all costs I must be myself.'

Neptune

Neptune is the opposite of Uranus: it represents the principle of universality and idealism. Expressed as a moral principle Neptune says: 'I am a member of the human race, and share a common destiny with my fellow men and women: I must therefore try as far as possible to break down the boundaries between myself and other people, and to submerge my own individuality in order to realize my wider humanity. I must try to empathize with other people, and to adopt a caring attitude towards them, irrespective of my personal whims. At all costs I must avoid being led astray by ideas and goals which are purely selfish and self-centred.'

Pluto

Pluto is the principle of dedication and single-mindedness. Expressed as a moral principle, Pluto says: 'Whatever I have decided to achieve, I must not rest until I have achieved it. I must be totally ruthless with myself in forcing myself onwards, and also totally ruthless and uncompromising in my dealings with the outside world. At all costs I must avoid apathy and inertia.'

The first four of these outer planets are clearly related to each other in a system of opposition and complementarity. Jupiter is the opposite of Saturn, and Uranus is the opposite of Neptune; but each of the other pairs of planets shares a quality which is missing from the opposite pair of planets. Thus, Jupiter and Uranus share a quality which we can call self-centredness, in which the aim is to enhance the self, whereas Saturn and Neptune share the opposite quality of other-centredness in which the aim is to limit the self and reduce its power. But, on the other hand, Jupiter and Neptune share a quality which we can call openness, in which the aim is to destroy or blur the boundaries between self and other and to be receptive to new influences; whereas Saturn and Uranus share the opposite quality of closedness, in which the aim is to strengthen the boundaries between self and other so as to prevent the infiltration of new influences. Diagrammatically, this can be represented as follows:

	Open	*Closed*
Self-centred	Jupiter	Uranus
Other-centred	Neptune	Saturn

Pluto stands outside this system, having qualities which are compatible with any of the qualities of the other planets (for one can pursue the aims of any of the planets in a dedicated and single-minded way). The isolation of Pluto within the scheme suggests that there may possibly be three undiscovered planets beyond Pluto, which are related to Pluto in the same way that the four previous planets are related to one another. (A planet which was the opposite of Pluto would say: 'I must cultivate detachment, passivity and non-involvement; I must always be prepared to compromise and to change my mind; I must at all costs avoid fanaticism.')

Nodes of the Moon

The absence of the Nodes from this book is sure to cause comment, and so I should explain their exclusion. I have omitted the Nodes for two reasons: firstly, because of the great confusion and disagreement in the astrological literature about how the Nodes should be interpreted; and secondly, because of the uncertainty about whether the Mean Node or the True Node should be used. The True Node can differ from the Mean Node by as much as 2 degrees, which becomes 18 degrees in the H9 chart; hence the difference between them is very significant in harmonic charts. In order to include the Nodes, I would have had to do research into whether the Mean Node or the True Node was more significant; and this would have been difficult, because of the uncertainty about how the Nodes should be interpreted. (Another problem has been that I have had no access to an ephemeris showing the True Node for birthdates before 1900.)

Hence I have excluded the Nodes; but readers are of course welcome to include the Nodes in harmonic charts, and to draw their own conclusions about their significance. One thing should be remembered, however. Despite the methods of astrologers such as Ebertin, the Nodes are *not* planets and should not be interpreted in the same way as planets. They are more likely to resemble the Ascendant and Descendant or the M.C. and the I.C.: that is, they can be seen as another pair of Angles to the chart. In an earlier publication I said:

> A planet which is aspected by another planet is thereby integrated with that planet: if it is aspected by a number of planets, it forms part of an integrated pattern. But a planet which is aspected by Ascendant or M.C. is not thereby integrated with anything: it is, rather, cast into prominence, brought out of darkness into light . . . .[2]

This is an essential difference.

Vertex and asteroids

The case for the inclusion of the vertex-antivertex axis in harmonic charts is perhaps stronger than the case for including the Nodes, and those astrologers who use the vertex will doubtless wish to enter it in harmonic charts.

Also, there is a strong case for including the asteroids. I believe myself that at least two of the asteroids (Ceres and Vesta) can usefully be included in the birth chart, and there is every reason to believe that their inclusion in harmonic charts would lead to interesting findings.

CHAPTER 10
ASPECTS BETWEEN PLANETS

In this chapter I will suggest interpretations for aspects between each pair of planets in each of the four harmonics H4, H5, H7 and H9. As I explained in Chapter 8, these should only be regarded as starting-points for interpretation; the precise interpretation in a particular case should depend on the houses and signs occupied in the radical chart, and on the other aspects which are linked to this aspect. In general, the closer an aspect is to exactitude, the stronger its effect will be.

The interpretations given below are especially descriptive of the oppositions in the H4 chart, and of the conjunctions in the H5, H7 and H9 charts. Readers must make their own adjustments to fit the interpretations to the other aspects. For instance, trines in the H4 chart can be described in terms of '*pleasure* in effort . . .', and squares in the H5 chart can be described in terms of '*striving* to adopt a style . . .'.

Sun-Moon

Sun-Moon contacts are especially important, since the Sun and the Moon are the two centres of the person's consciousness or awareness. If the Sun and the Moon come together in a particular harmonic, the person seems to carry the qualities of that harmonic as a general 'aura', which colours his personality and envelops all of his behaviour, even if the harmonic chart contains no other important aspects.

H4 Striving to find oneself: restless, dynamic, tense, uneasy: tending to be dissatisfied with present circumstances: seeing life as a challenge: needing constant fresh challenges.

H5 Living one's life in an orderly, patterned way: preoccupied with style, and with self-image: seemingly stylized and artificial behaviour: a creature of habit.

H7 Having a romantic vision of oneself and the world: seeming to others to be a romantic and inspiring person: impulsive, unpredictable: having a sense of one's mission to inspire others.

H9 At peace with oneself: experiencing joy in oneself and in one's relationship with the world: innocent, naive, childlike: having the confidence (through lack of fear) to engage in ambitious and daring behaviour.

Sun-Mercury

Sun-Mercury contacts in the harmonic charts are particularly interesting, since in the radical chart no aspects between Sun and Mercury are possible except the conjunction. The theme of Sun-Mercury aspects is 'thinking and communicating about oneself'.

H4 Engaged on a search for self-understanding, or a voyage of intellectual enquiry about oneself: puzzling about one's own nature: unsure of one's opinions: searching for methods of self-expression: expressing opinions assertively so as to cover uncertainty.

H5 Thinking about oneself and one's environment in orderly, patterned ways: building elaborate thought-structures around oneself: communicating about self and opinions in terms of these structures.

H7 Thinking and communicating about oneself and one's environment romantically and inspirationally: having fragmented, ephemeral, fantastical self-theories: great verbal facility (especially with the trine).

H9 Finding joy and peace from thinking and communicating about oneself and one's place in the world: intellectually contented and complacent.

(The following aspects are never found, except where the conjunction is carried over from the radical chart:
H4 the conjunction, opposition, and trine
H5 the conjunction and opposition
H7 and H9 the conjunction.)

Sun-Venus

H4 Aware of a lack of self-love, and striving to rectify this: needing to be reassured by the appreciation, admiration and love of other people: attempting in an 'edgy' way to charm and attract other people: judgemental about other people: 'keeping oneself to oneself' in the presence of people whom one dislikes.

H5 Developing styles of behaviour which will attract other people and make one feel comfortable in their presence: very anxious to please: tending (if one cannot find a style of behaviour to fit the occasion) to become withdrawn and taciturn.

H7 Developing romantic feelings about oneself and one's place in the world: seeking to attract others through romantic and inspirational behaviour: tending to have wild and unrealistic value-systems.

H9 Obtaining joy and peace from appreciation of self: rejoicing in one's ability to please and attract other people: tending to complacency about one's attractiveness. (A common aspect among actors and performers with sex appeal.)

(The conjunction is never found in the H4, H5 and H7 charts, except when carried over from the radical chart.)

Sun-Mars

H4 Striving to demonstrate to the world one's energy and forcefulness: striving to bend others to one's will: striving to be manly and virile: a tendency to exhibitionism, especially with the 'soft' aspects.

H5 Developing a style of forceful action through which one can express one's personality: being a 'stylist' in everything that one does.

H7 Able to express one's forcefulness in inspirational and impulsive ways: inspired by an ideal of sexual forcefulness, or by a sublimation of this, e.g. military valour or sporting prowess: often great physical magnetism. (The effects of this aspect are always striking.)

H9 Finding joy in one's own physicality, sexuality and forcefulness (but often seems to be a 'recessive' aspect, with no visible effects).

Sun-Jupiter

H4 Striving to overcome the problems and difficulties in the path of self-development: struggling to break free from restrictions and to grow to one's full potential: struggling to develop oneself as a leader of men (a common aspect among politicians): with the 'soft' aspects, a tendency to wry and self-deprecatory humour.

H5 Finding a style of behaviour within which one can develop one's own freedom and growth: projecting a stylized image of oneself as a free, uncluttered, 'larger than life' personality. (A common aspect among actors.)

H7 Inspired by the ideals of growth and freedom: developing a philosophy of freedom which inspires one's actions. (A common aspect among writers with a philosophical bent.)

H9 Delighting in the experience of freedom: rejoicing in the feeling that there is nothing that one cannot do: able to help others to grow. (A common aspect among spiritual leaders.)

Sun-Saturn

H4 Striving to introduce control and self-discipline into one's life: struggling to be realistic in one's self-expectations and to purge oneself of vain ambition: having a tendency to opt out of things that bring pleasure and to struggle for things that bring pain or hardship: tending to be distrustful of other people.

H5 Developing self-restricting styles of behaviour, which help one to come to terms with one's limitations and one's destiny: projecting an image of oneself as a controlled and down-to-earth person.

H7 Inspired by ideals of duty, dedication, orderliness, loyalty, acceptance of fate: believing in the necessity of suffering and sorrow: seeing oneself as a tragic person.

H9 Rejoicing in one's down-to-earth practicality and realistic and commonsense attitude to life: an anti-religious aspect.

Sun-Uranus

H4 Striving to establish oneself as an original personality: rejecting mediocrity in every form: striving to shock and disturb other people out of complacency: a common aspect among dictators and among eccentrics of all kinds.

H5 Developing precise, sharp, clearly-defined styles of behaviour which emphasize one's difference from other people: rejecting 'fuzziness' and ambiguity.

H7 Inspired by ideals of individuality, originality, clarity and brilliance: having a highly individual vision: able to shock and disturb other people by one's inventive fantasies.

H9 Revelling in one's individuality and one's difference from other people: going one's own way, regardless of the views of others.

Sun-Neptune

H4 Striving to lose one's individuality and to emphasize shared identity with larger groups of people or with the whole of mankind: striving to care about other people and to feel that one's own problems are theirs.

H5 Developing styles of behaviour which emphasize one's sensitivity, caringness, concern for humanity and lack of concern for self.

H7 Inspired by ideals of service to others: inspired by the idea of sinking one's individuality in a wider cause: seeing oneself as a symbol of something greater than oneself. (The *square* aspect seems to be very productive of creative genius.)

H9 Rejoicing in the knowledge of one's own insignificance: rejoicing in one's ability to serve the community: a very selfless aspect.

Sun-Pluto

H4 Striving to be ruthless with oneself: refusing ever to rest on one's laurels, since one always feels that one should drive oneself still further: having a sombre and serious approach to life.

H5 Developing self-punishing styles of behaviour: working relentlessly: denying expression of the more carefree sides of one's personality.

H7 Inspired by ideals of self-punishment and self-transformation: exaggerating, or having fantasies about, one's own faults and defects: favouring strenuous and self-punishing forms of therapy.

H9 Rejoicing in self-punishment and self-belittlement: rejoicing in revealing oneself to the world 'warts and all'.

Moon-Mercury

H4 Effort to articulate and communicate one's feelings and perceptions: effort to make sense of one's environment: above all, these aspects are concerned with the *search for language* with which to express what at first seems incommunicable.

H5 Adopting a style of intelligent perceptiveness and of articulate communication of feelings: creating a rational structure of thought for understanding the world: seems to be connected with wit, humour, and especially with satire.

H7 Easily moved and inspired by the idea of articulateness, mental adaptability, open-mindedness: having an experimental outlook: mental exploration.

H9 Joy in communication of perceptions and in articulation of feelings: happiness in the feeling that one understands the nature of the world.

Moon-Venus

H4 Effort to find beauty (love, tenderness, gracefulness) in a world that seems to lack these things: effort to respond aesthetically: effort to develop the feminine side of one's nature, and to attract love and appreciation.

H5 Responding to beauty of form and structure: adopting a graceful, sensitive and attractive style.

H7 Inspired by the idea of awareness of beauty, and of femininity. Seemingly can lead (in men) either to effeminacy or to inconstancy with women.

H9 Joy in the perception of beauty (love, tenderness) in the environment: seeing the world as a beautiful place: rejoicing in the femininity of one's behaviour, and in one's ability to attract.

Moon-Mars

H4 Effort to overcome the inertia which seems to surround one, and to arouse others to action: effort to find actions which express one's feelings: effort to respond to events dynamically and aggressively: effort to change the environment by force.

H5 Adopting a style of quick, dynamic and forceful response to events, and of open, direct and powerful communication: may be connected with knock-about humour.

H7 Inspired by the idea of dynamic and forceful response to events: emotionally affected by whatever is active, forceful, dynamic: tendency to dramatic and impulsive gestures, and to exhibitionism: sometimes an exaggerated sexuality.

H9 Joy in being energized by the environment and by other people: seeing the world as full of dynamic action: rejoicing in being able to change the world through one's actions.

Moon-Jupiter

H4 Effort to respond to people and to events in an open, impressive, uninhibited way: effort to overcome the apparent 'pettiness' of one's environment, and to invest it with meaning and grandeur: effort to make the most of one's opportunities: effort to free the world from restraint.

H5 Adopting a style which involves free and open expression of feelings: instinctive over-acting and self-dramatization.

H7 Inspired by the idea of grand and uninhibited response to events: emotionally affected by whatever is grand and noble and expressive: a very creative aspect, often leading to art which is full of 'grand emotion'.

H9 Joy in being uplifted by one's environment: joy in responding to opportunities for growth and development: instinctive feelings of freedom and optimism.

Moon-Saturn

H4 Effort to respond to events in a controlled and disciplined way: effort to limit one's freedom of response, and to follow the path of duty.

H5 Adopting a style which involves strict control of feelings: adopting a 'hard' attitude to the world, with deliberate lack of sensitivity: dry humour: building defensive structures against the environment.

H7 Inspired by the idea of self-control: emotionally affected by sadness, misfortune, oppression, devotion to duty: can lead to self-pity and to impulsive acts of self-abnegation.

H9 Joy in the acceptance of difficulties and constraints: joy in accepting the world as it is, without illusions: cheerful cynicism.

Moon-Uranus

H4 Effort to respond to events in a vivid, sharply-defined, and individualistic way: effort to overcome the dullness and confusion of the environment, and to give it clarity and excitement: effort to stamp one's personality on the environment.

H5 Adopting a style of vivid and electrifying response to events, with assertion of one's originality: very conducive to the development of new, dynamic and sometimes 'shocking' styles in art, music, literature, acting or politics.

H7 Inspired by the idea of vivid and electrifying reponse to events: emotionally affected by whatever is striking, original, startling: can lead to great daring and foolhardiness, and to a 'couldn't-care-less' attitude to other people.

H9 Rejoicing in the vividness and excitement of the world: childlike enthusiasm.

Moon-Neptune

H4 Effort to respond to events in a caring and idealistic way: effort to transcend one's own boundaries, to come closer to other people, and to feel a part of one's environment.

H5 Adopting a style of response which emphasizes universal values rather than individual differences: seeing the world in terms of idealistic systems, structures and codes.

H7 Inspired by an idealistic vision of humanity: emotionally affected by things that are symbolic of the human condition: dreamily (sometimes sentimentally) romantic.

H9 Instinctively perceiving goodness or spirituality in others: having a sense of the mystic unity of the universe (pantheism): rejoicing in one's ability to transcend one's own boundaries and to come closer to other people and the environment.

Moon-Pluto

H4 Effort to respond to events in a ruthless, self-punishing way: effort to dig deep into one's emotions and bring them to the surface.

H5 Adopting a hard, ruthless, self-punishing style of behaviour in response to events. In some cases the person seems to be bent on self-destruction (physical or mental) through his behaviour: in other cases there is a facade of hardness and ruthlessness.

H7 Inspired by the idea of ruthlessly self-punishing energy, of transformation and of death and re-birth: emotionally affected by whatever is profound, secret, awesome: ruthless in seeking inspiration from one's subconscious mind.

H9 Rejoicing in the awesomeness of the universe: rejoicing in one's ability to press oneself to the limit in response to events.

Mercury-Venus

H4 Effort to communicate charmingly and persuasively: effort to close the gap between thought and feeling, and to achieve beauty of thought and rational values.

H5 Adopting a stylized charm: finding stylized ways of communicating wittily, charmingly and persuasively, and of communicating the feminine side of one's nature: building beautiful thought-structures or rational theories of beauty.

H7 Inspired by the idea of beauty in rationality, and of charming and delightful communication: having an Arcadian vision: having an impulsive charm.

H9 Rejoicing in the beauty of one's thoughts, in the rationality of one's values, and in one's ability to communicate charmingly and persuasively.

(The conjunction in the H4 chart cannot occur, except when carried over from the radical chart. The conjunction in the H5 chart is rare – except when carried over – since both Mercury and Venus need to be at maximum elongation on opposite sides of the Sun.)

Mercury-Mars

H4 Struggle to achieve mental forcefulness and decisiveness: effort to close the gap between thought and action, and to think positively and act rationally: effort to have the courage of one's convictions.

H5 Adopting a forceful, dynamic, rugged, aggressive style of communication: masculine patterns of thought: having positive thought-systems that determine one's actions.

H7 Inspired by the idea of a rational approach to action and sexuality, and by the idea of forceful, dynamic, rugged communication: impulsively communicating with great force.

H9 Rejoicing in one's ability to think positively, communicate forcefully, and act rationally: rejoicing in one's ability to achieve goals, and to convert thoughts into action.

Mercury-Jupiter

H4 Struggle to achieve mental freedom and growth: effort to free the mind from restrictive thought-patterns: effort to communicate in an uplifting and impressive way: wanting to be admired for one's open and enquiring mind.

H5 Adopting an open style of communication, with a tendency towards free association and lateral thinking: having thought-systems that centre around ideas of freedom and growth: finding stylized ways of communicating one's impressiveness and one's freedom from restraint: talent as an educator.

H7 Inspired by the idea of free and uninhibited thought and communication, leading to freedom and growth: free-ranging imagination: inspired to generosity and magnanimity: prone to exaggeration in communication (purple passages, ham acting, overstating one's case).

H9 Delighting in one's open-mindedness and freedom from restrictive thought-patterns: rejoicing in one's ability to spread good cheer through communication.

Mercury-Saturn

H4 Struggle to acquire mental discipline: effort to channel the mind, and to tie it down within restrictive thought-patterns: effort to direct the mind to practical affairs, and to communicate effectively at a practical level: wanting to be admired for one's common sense and realism.

H5 Adopting a style of controlled and disciplined thought and communication: having thought-systems that centre around practical reality and the absence of illusion: finding stylized ways of communicating one's earthiness and practicality: can lead to great articulateness: can also provide the mental discipline which leads to success in various creative fields.

H7 Inspired by the idea of controlling and disciplining the mind, and of thinking and communicating in a down-to-earth and practical way: inspired by the idea of scientific proof and experimentation, and of belief systems that limit one's freedom of thought: inspired to communicate one's limitations and obligations, and those of other people.

H9 Rejoicing in the controlled and disciplined nature of one's thoughts, and in one's mental realism and practicality: rejoicing in being able to communicate effectively: rejoicing in the absence of illusion.

Mercury-Uranus

H4 Struggle to achieve mental clarity and originality: effort to communicate in original, stimulating and striking ways that demonstrate the essence of one's individuality.

H5 Adopting a precise and individualistic style of thought and communication: having thought-systems that emphasize one's originality and separateness from other people: communicating in a 'clipped' and precise fashion (with a tendency to taciturnity, since there may be an obsessive desire to avoid clichés or accepted patterns of thought).

H7 Inspired by the idea of original, striking, electrifying forms of thought and communication: mental originality at all costs: liking to shock by the way that one communicates: inspired by belief-systems that emphasize the uniqueness of the individual: talent as an inventor.

H9 Rejoicing in the clarity and originality of one's mind and in the ability to communicate one's individuality to others.

Mercury-Neptune

H4 Struggle to close the mental gap between oneself and other people and communicate meaningfully to others: effort to think and say things of universal significance: wanting to be admired for one's capacity to think imaginatively: struggle to become a prophet.

H5 Adopting a sympathetic and caring style of thought and

communication: having thought-systems that emphasize one's common humanity and de-emphasize one's individuality: conducive to the creation of poetry and imaginative fiction.

H7 Inspired by the idea of communicating words and thoughts of universal significance, or by the idea of being a prophet: inspired by belief-systems that emphasize the oneness of mankind.

H9 Rejoicing in one's ability to communicate in such a way as to 'touch' other people and make life happier for them: rejoicing in one's mental empathy with other people.

Mercury-Pluto

H4 Struggle to pursue one's thoughts to their ultimate logical conclusion: effort to communicate relentlessly and powerfully: effort to overcome mental flabbiness and inconsistency, and to be admired for the power of one's intellect: struggle to reveal all about oneself.

H5 Adopting a self-denying style of thought and communication, often with a self-deprecatory style of humour: having thought-systems that emphasize the insignificance of the individual and the need for self-castigation and for relentless mental routines.

H7 Inspired by the idea of logical ruthlessness, or of allowing reason to outweigh all other considerations: inspired by ruthless and harsh belief-systems.

H9 Rejoicing in one's ability to communicate one's thoughts relentlessly, fearlessly and consistently, without regard for personal consequences: rejoicing in the impersonal truth.

Venus-Mars

H4 Struggle to unite the male and female sides of one's nature: struggle to express one's values forcefully: struggle to find a course of action which is true to one's values: struggle to be loved for one's strength and dynamism: the search for beauty in action: in a woman, effort to be strong without losing one's femininity: in a man, effort to be attractive and loving without losing one's masculinity.

H5 Adopting a style which combines masculine force with feminine persuasiveness, or masculine strength and vigour with feminine grace and beauty: a very seductive aspect.

H7 Romantically in love with the idea of power, force and dynamic action: loving to be loved for one's forcefulness, dynamism or sexuality.

H9 Rejoicing in one's ability to charm people by one's forcefulness and dynamism: having an internal radiance which results from lack of conflict between one's masculine and feminine natures.

Venus-Jupiter

H4 Striving to love freedom, and to be loved for one's freeness, open-mindedness, impressiveness or grandeur: striving to attract others by grand and impressive behaviour: seeking to sublimate the search for love: effort to free the mind from restrictive value-systems and restrictive patterns of enjoyment: the search for a sublime beauty.

H5 Finding stylized ways of expressing a love of freedom: having value-systems that centre around the love of freedom and growth: able to attract others by one's openness and generosity: a love of free and unrestrained forms in art.

H7 Romantically moved by ideas of freedom, growth and progress: believing in uninhibited and unconditional love: inspired to grand and extravagant displays of feeling: seeing beauty in whatever is grand, expansive and uplifting: a romantic optimist.

H9 Delighting in one's ability to express one's beliefs and values openly and expansively, and to affect others with one's joy and optimism: attracting others with one's qualities of warmth, openness and generosity.

Venus-Saturn

H4 Striving to love duty: striving to attract others through one's self-control, faithfulness, conscientiousness and devotion to duty: seeking to be realistic in the search for love, and to curb one's wilder expectations: effort to tie the mind down to realistic and practical value-systems: the search for beauty in down-to-earth, everyday things.

H5 Finding stylized ways of expressing a love of duty: having value-systems that centre around discipline, control and restraint: able to attract others by one's earthiness and devotion to duty: a love of restrained and disciplined forms in art.

H7 Romantically moved by ideas of suffering, hardship and devotion to duty: a belief in love attained through tribulation: inspired to displays of suffering: seeing beauty in whatever is sad, tragic, expressive of earthly tribulations: a romantic pessimist.

H9 Delighting in one's ability to express one's pleasure in ordinary earthly things, and to spread this pleasure to others so as to increase their satisfaction with their earthly lot: attracting others with one's qualities of solidity and earthiness.

Venus-Uranus

H4 Striving to love originality: striving to attract others by one's uniqueness, vivacity and unpredictability: striving to repel those who do not accept one for these qualities: striving for adventure in love, and for unconventional and sharply-defined value systems: tending to withdrawn-ness and taciturnity if these expectations are not met: the search for a vivid and thrilling beauty.

H5 Finding stylized ways of expressing a love of clarity and originality: having value-systems that centre around innovation and adventure: having a vivacious charm, able to attract others by one's sharply-defined individuality: loving original, exciting and sharply-defined forms in art.

H7 Romantically moved by ideas of adventure, excitement, innovation, brilliance: believing in perfect love (i.e. love that is sharply-defined, unique, unmarred by defects, lightening the world with its brilliance): inspired to sudden and electrifying displays of feeling: seeing beauty in whatever is new, original, innovative and brilliant.

H9 Delighting in one's ability to express one's beliefs and values in a vivid, exciting and individual way, and to affect others with this excitement: attracting others with one's qualities of uniqueness, magnetism, unpredictability.

Venus-Neptune

H4 Striving to love all mankind: striving to attract others through one's sensitivity and awareness of the human condition: wanting to be loved as a symbol of something greater than oneself: striving for unselfish love: the search for universal value-systems and for an all-embracing beauty.

H5 Finding stylized ways of expressing a universal love of humanity or sympathy for the human condition: having value-systems that centre around the submergence of individuality in the greater good: developing an elusive charm, able to attract others by one's air of caringness and sensitivity: a love of flowing forms in art, with no sharp boundaries or vivid effects.

H7 Romantically moved by ideas of caring, service, improving the lot of mankind: a belief in selfless and universal love: inspired to displays of feeling about universal issues: seeing beauty in whatever is mystical, symbolic, universal in its implications: very deeply romantic.

H9 Delighting in one's ability to empathize with the feelings of others, and to express one's beliefs and values in a way which carries meaning for others: attracting others with one's qualities of sensitivity and mysteriousness.

Venus-Pluto

H4 Striving to love dedication: striving to attract others through one's ruthless determination and single-mindedness: seeking a fierce and devouring love: the search for ruthless and self-denying value-systems, and for a 'terrible beauty'.

H5 Finding stylized ways of expressing a love of dedication and single-mindedness: having value-systems that emphasize the single-minded pursuit of a goal: able to attract others by one's dedication and ruthlessness in pushing one's abilities to the limit: loving consistent and uncompromising forms in art.

H7 Romantically moved by ideas of dedication, ruthlessness, transformation, death and re-birth: believing in fanatical and all-consuming love: inspired to displays of strong and impassioned feeling: seeing beauty in that which penetrates below the surface to the deepest meaning.

H9 Delighting in one's ability to express one's beliefs and values ruthlessly and unflinchingly, so as to affect and change the values of others: attracting others with one's qualities of unflinching dedication.

Mars-Jupiter

H4 Striving towards free and uninhibited action, and the free and unrestrained expression of physicality and sexuality: striving to free one's behaviour from rules and restraints, and to make decisions which will lead to personal growth and development.

H5 Finding stylized ways of expressing a love of freedom in action: developing skill in activities which involve a display of physical impressiveness, daring and independence: in art, preferring grand and uninhibited gestures and free expression.

H7 Inspired by the uninhibited expression of physicality and sexuality: inspired by grand projects and noble aims: believing that 'man should be free, when everywhere he is in chains': expressing this message through one's actions and one's art.

H9 Rejoicing in one's capacity for optimistic action and for making the most of opportunities: delighting in one's ability to affect others with one's optimism and force them to adopt a more positive approach.

Mars-Saturn

H4 Striving towards a code of conduct which will govern one's actions and limit one's physical and sexual freedom: the effort to act in a controlled and disciplined way in order to achieve laid-down objectives: striving to do one's duty.

H5 Finding stylized ways of expressing a love of controlled and disciplined action: developing skill in activities which involve the channelling of physical effort and the determined application of a set of rules. (A common aspect among astrologers.)

H7 Inspired by the idea of doggedly pursuing the path which destiny has laid down for one: believing that men's wilder instincts should be restrained: expressing this message through one's actions and one's art.

H9 Rejoicing in one's capacity for dogged endeavour and for accepting setbacks and cheerfully coping with difficulties: rejoicing in acting out one's allotted role.

Mars-Uranus

H4 Striving to act in original, dramatic and magnetic ways, and in ways that are completely expressive of one's own individuality: striving towards physical and sexual adventure: striving to free one's behaviour from dull routines, and to be free to act unpredictably on the spur of the moment.

H5 Finding stylized ways of expressing a love of vivid and electrifying action: developing skill in activities which call for a completely individualistic approach, and for the display of sheer physical energy.

H7 Inspired by the idea of acting in original and startling ways that owe nothing to convention: believing in being completely free to express one's individuality: expressing this message through one's actions and one's art.

H9 Rejoicing in one's ability to act in original and independent ways: rejoicing in one's physical uniqueness: able to affect others with one's enthusiasm and joy of life.

Mars-Neptune

H4 Striving to find idealistic goals for one's actions: striving to escape from the pointlessness of everyday life, and to feel that one is serving a cause: wanting to be feared as a symbol of something greater than oneself.

H5 Finding stylized ways of expressing a love of idealistic and sympathetic action: developing skill in activities which involve subordinating one's individuality to a common cause.

H7 Inspired by actions which serve the cause of humanity: believing in idealistic programmes of action, and in the subordination of one's individuality to a wider cause: expressing this message in one's actions and one's art.

H9 Rejoicing in one's ability to act in accordance with one's ideals and to serve others rather than oneself: rejoicing in one's ability to affect others with one's idealism.

Mars-Pluto

H4 Striving to push one's physical resources to their limits: striving to work relentlessly towards one's goals: wanting to be feared for one's ruthlessness: striving to overcome apathy and inertia.

H5 Finding stylized ways of expressing a love of ruthless determination in action: developing skill in activities which require total dedication and relentless self-punishment in order to achieve success.

H7 Inspired by programmes of action that are ruthless, hard and uncompromising: believing in driving oneself very hard to achieve results: expressing this message in one's actions and one's art.

H9 Rejoicing in one's capacity for hard work and ability to push one's physical resources to their limits: rejoicing in one's ability to affect others with one's dedication.

Jupiter-Saturn

H4 Striving to resolve a dilemma between discipline and freedom: striving towards a type of freedom which is contained within bounds, or recognizes its limitations.

H5 Having a style which combines expansiveness with restraint: able to move expansively towards controlled expression, or in a controlled way towards expansive expression.

H7 Inspired by the combination of grandeur and nobility with restraint and suffering: able to express this creatively, depending on other aspects.

H9 Rejoicing in the ability to combine freedom with acceptance of one's fate, and to achieve freedom within one's prescribed role.

Jupiter-Uranus

H4 Striving to resolve a dilemma between freedom and originality (i.e. the fear that by growing towards freedom, one may lose separateness and uniqueness; or that, by defining one's originality, one may lose freedom and be unable to grow): striving towards an idiosyncratic type of freedom: striving to increase one's receptivity to new ideas.

H5 Having a style which combines expansiveness and 'free form' with sharpness, clarity and originality.

H7 Inspired by the combination of grandeur and nobility with excitement, brilliance and shockingness: able to express this creatively, depending on other aspects.

H9 Rejoicing in one's ability to combine freedom and optimism with the realization of one's separate individuality, and to achieve a type of freedom which is unique to oneself.

Jupiter-Neptune

H4 Striving to resolve a dilemma between freedom and connectedness (i.e. the fear that by moving closer to others one may lose personal freedom, or that by choosing the path of personal growth one may cut oneself off from others): striving towards an idealistic type of freedom.

H5 Having a style which combines expansiveness with sympathy and caringness: able to express universal significance in an expansive way.

H7 Inspired by the combination of grandeur and nobility with idealism, caringness, and mystical significance: able to express this creatively, depending on other aspects.

H9 Rejoicing in one's ability to combine freedom and optimism with a feeling of empathy with mankind, and to achieve freedom through the dissolution of the boundaries of the self.

Jupiter-Pluto

H4 Striving to resolve a dilemma between freedom and single-mindedness (i.e. the fear that by choosing freedom and open-mindedness, one may lose the ability to achieve goals by single-minded dedication, or vice versa): striving towards the relentless pursuit of growth and freedom.

H5 Having a style which combines expansiveness with ruthlessness.

H7 Inspired by the combination of grandeur and nobility with ruthless exposure of oneself and of others: able to express this creatively, depending on other aspects.

H9 Rejoicing in one's ability to combine freedom with dedication, and to achieve freedom through self-punishment.

Saturn-Uranus

H4 Striving to resolve a dilemma between discipline and independence, or between doing one's duty and 'doing one's own thing': striving towards a disciplined originality, or a self-imposed discipline.

H5 Having a style which combines sharpness, clarity and originality with control and restraint: able to express one's uniqueness in a restrained or down-to-earth way.

H7 Inspired by the combination of excitement and brilliance with control and restraint: inspired by the idea of controlled and bounded originality: able to express this creatively, depending on other aspects. (The list of famous people who have this aspect is unusually long, and it would seem that it is a powerful stimulus to ambition.)

H9 Rejoicing in the ability to realize one's individuality fully within the constraints imposed by destiny, and to reconcile originality with the performance of duty.

Saturn-Neptune

H4 Striving to resolve a dilemma between discipline and connectedness (i.e. the fear that by caring about others one may lose sight of one's destiny, or that by doing one's duty one may lose sight of the needs of others): striving towards selflessness and a disciplined service to the needs of others.

H5 Having a style which combines control, restraint and earthiness with sympathy and caringness: able to present oneself as a selfless person who gives priority to the needs of others.

H7 Inspired by the combination of control and restraint with idealism, caringness and mystical significance: inspired by the idea of self-abnegation in the cause of humanity: able to express this creatively, depending on other aspects.

H9 Rejoicing in the feeling that, in doing one's duty or accepting one's limitations, one is serving mankind: rejoicing in the selflessness of one's motivation.

Saturn-Pluto

H4 Striving to resolve a dilemma between duty and single-mindedness (i.e. between doing one's duty as defined by others, and the ruthless pursuit of one's own goals): striving towards a disciplined single-mindedness.

H5 Having a style which combines control, restraint and earthiness with ruthlessness and single-minded consistency: ruthlessly developing a disciplined style of behaviour.

H7 Inspired by the combination of control and restraint with ruthlessness and dedication: inspired by the idea of single-minded devotion to duty: able to express this creatively, depending on other aspects: possibly connected with sadism and masochism.

H9 Rejoicing in one's single-minded devotion to duty and realistic acceptance of limitations: delighting in deprivation.

Uranus-Neptune

H4 Striving to resolve a dilemma between self-centredness and selflessness, or between separateness and connectedness: striving to serve the needs of others in an individualistic way, or to make one's uniqueness a symbol of a wider truth.

H5 Having a style which combines clarity and originality with outgoingness and caringness: a style which conveys the message 'I am in my separate individuality but am willing to meet you in yours': alternatively a style whose purpose is to invest one's individuality with universal significance.

H7 Inspired by the combination of excitement and brilliance with caringness and mystical significance: inspired by the idea of the individual becoming a symbol of a wider truth: a very creative aspect.

H9 Rejoicing in being simultaneously an individual and a member of the human race: rejoicing in serving the cause of humanity by one's independent actions.

Uranus-Pluto

H4 Striving to resolve a dilemma between originality and single-mindedness, or between the desire to be unpredictable and the desire to be consistent: seeking ruthless ways of establishing uniqueness, or original ways of achieving goals.

H5 Having a style which combines clarity and originality with ruthless consistency: ruthlessly developing an original style of behaviour.

H7 Inspired by the idea of ruthless dedication to the cause of one's unique individuality: able to express this creatively, depending on other aspects.

H9 Rejoicing in the feeling that, through total dedication, one is able to realize fully one's individual potential.

Neptune-Pluto

H4 Striving to resolve a dilemma between connectedness and dedication (i.e. the fear that sympathy for others may deflect one from the path, or that one's single-mindedness may cut one off from other people): seeking ruthless ways of achieving selfless goals.

H5 Having a style which combines ruthless consistency with a concern for the needs of others, and a concern for universal truth and mystical significance.

H7 Inspired by the idea of ruthless dedication to the pursuit of universal truth or to the cause of humanity: able to express this creatively, depending on other aspects.

H9 Rejoicing in the feeling that, through total dedication, one can break through into a higher truth and a more universal level of understanding.

CHAPTER 11
ASPECTS TO THE ANGLES

As we have already said (Chapter 3), the Angles in harmonic charts must be treated with great caution, because of the inaccuracy of most birth times. Any inaccuracy in the radical chart is multiplied in the harmonic chart by the number of the harmonic. If the given birth time is ten minutes away from the true birth time, then the Midheaven will be wrong by 2½ degrees: this may not seem like a large error, but in the H9 chart it will be magnified to 22½ degrees, which is enough to completely invalidate any conclusions about planets forming aspects to the Angles.

Because of this, I have not been able to do any systematic study of the Angles in harmonic charts. But, despite this, I believe that aspects to the Angles are important; and in fact their importance is often clearly shown in cases where the birth time is known with reasonable accuracy. I would suggest, therefore, that aspects to the Angles should be included in the interpretation in cases where the birth time is known to be accurate to within five minutes.

In the case of the H5, H7 and H9 charts, I note only *conjunctions* to the four Angles (Ascendant, Descendant, M.C., I.C.), using an orb of 12 degrees (the same as is used for conjunctions between planets). A note can be taken of other aspects (using harmonic orbs: 4 degrees for the trine, 3 degrees for the square, and so on), but in my opinion these aspects are best ignored unless the time of birth is *very* accurately known.

The case of the H4 chart is different. Because it is an even-numbered harmonic chart, each point on the zodiac is united with its opposite point, so that the Ascendant is also the Descendant and the M.C. is also the I.C. The point which is opposite to the Ascendant in the H4 chart is not the Descendant, but is the point opposed to the Ascendant/Descendant axis (this axis having collapsed to a point). Therefore, if we use an orb of 12 degrees for conjunctions to the Ascendant/Descendant, we should use an orb of only 6 degrees for oppositions to this point. However, we may also (if the birth time is accurate enough) note planets which are square to the Ascendant/Descendant, using an orb of 3 degrees. (The same applies for aspects to the M.C./I.C.)

In the remainder of this chapter I will make tentative suggestions for the interpretation of planets aspecting the Angles in harmonic charts. Because of the lack of accurate birth times, these suggestions are made more on the basis of theory than of empirical investigation, and they are therefore very sketchy and provisional. The interpretations are based on a particular view of the meaning of the Ascendant/Descendant and M.C./I.C. axes which I have found to be valid in my own interpretative work. This is as follows:

(a) The *Ascendant/Descendant axis* is concerned with one's relationship with other people and with the external world at the present time. People whose planets cluster around this axis tend to 'live in the present', and to be little concerned with the past or the future. The Ascendant is particularly concerned with how one presents oneself to the world, and the Descendant is particularly concerned with the view which one has of the outside world. However, both ends of the axis are basically concerned with the same thing, which is the process of interaction between self and other.

(b) The *M.C./I.C.* axis is concerned with the path which one follows through life, one's voyage out of the past and into the future. People whose planets cluster around this axis tend to be greatly involved with the past and the future, and to be relatively little concerned with the 'here and now'. The I.C. is particularly concerned with the past (origins) and the M.C. with the future (ultimate goals), but both ends of the axis are basically concerned with the same thing, which is the development of self through time.

H4 chart

Planets conjunct Ascendant/Descendant or M.C./I.C.

When a planet is conjunct an Angle in the H4 chart, it will have been either conjunct or square to an Angle in the radical chart. The interpretation should therefore be based on its position in the radical chart. (Note that when a planet is square to the Ascendant it is also square to the Descendant: both are equally important. The same applies to the M.C. and the I.C.)

Planets opposite or square to the Ascendant/Descendant

When a planet occupies this position in the H4 chart, the person feels that this force is *lacking* in his relationship with the outside world, and strives to overcome this lack. For example, if Mars occupies this position, the person will feel that his method of interacting with the world lacks forcefulness, decisiveness and dynamism, and he will strive to put this right. Hence there may seem to be a contrived forcefulness in his behaviour.

Planets opposite or square to the M.C./I.C.

When a planet occupies this position in the H4 chart, the person feels that the qualities of this planet are missing from the way in which he approaches his self-development, and he will strive to overcome this lack. For example, if Jupiter occupies this position, the person will feel that he needs to plan for the future in a more optimistic spirit, taking more risks and making better use of opportunities, and that he should also adopt a more positive approach to the talents and advantages which he brings with him from the past. He will do his best to achieve this, and so he will tend to plan his life in a spirit of deliberate optimism and risk-taking.

Sun Effort towards self-involvement: effort to be more involved in what one is doing.

Moon Effort to be more responsive to the environment and to other people: effort to respond to the needs of the moment.

Mercury Effort to be more rational, reasonable, communicative.

Venus Effort to act more in accordance with one's values, and to be more attractive (i.e. to behave so as to please other people).

Mars Effort to be more forceful, energetic, dynamic, decisive.

Jupiter Effort to be more optimistic, cheerful, risk-taking, working towards freedom and growth.

Saturn Effort to be more responsible, controlled, realistic, dutiful, down-to-earth.

Uranus Effort to be more original, stimulating, spontaneous, true to one's own nature, unconventional.

Neptune Effort to be more selfless, caring, sympathetic, idealistic, unemphatic.

Pluto Effort to be more consistent, powerful, single-minded, uncompromising.

H5 chart

Planets on the Ascendant/Descendant axis
Planets on this axis in the H5 chart will tend to be prominent within the stylized behaviour which the person adopts in his interactions with other people and with the world in general. For instance, if Uranus is on the Ascendant/Descendant axis in H5, the person will tend to adopt stylized patterns of behaviour which display on the surface the Uranian characteristics which are listed below. If there is an opposition between a planet on the Ascendant and a planet on the Descendant, then it *may* be that the rising planet represents the qualities which the person tends to display in his style, while the setting planet represents his perception of the qualities which other people display in their habitual manner of dealing with him.

Planets on the M.C./I.C. axis
Planets on this axis in the H5 chart will tend to represent the goals towards which the person is working within his stylized behaviour. They may represent talents or skills which he wishes to acquire, or behavioural attributes which he feels he must develop in ordered and structured ways, through the development of habits and routines. If there is an opposition between a planet on the M.C. and a planet on the I.C., then it *may* be that the planet on the I.C. represents the attributes which the person brings with him from the past, while the planet on the M.C. represents the attributes which he feels it is his destiny to develop in the future.

Sun Habit of self-involvement. (Sun on the Ascendant/Descendant axis in H5 seems to mean that the person needs to display himself – his true nature – to the world through his style, and so will decrease his ability to develop styles and mannerisms whose purpose is to deceive or to hide the self. Sun on the M.C./I.C. axis may decrease his *desire* to develop such styles.)

Moon Responsiveness (acting as a mirror to the world).

Mercury Intellectualism, theory-making, talkativeness, communicative skills.

Venus Stylized attractiveness, habitual judging and valuing.

Mars Stylized force and aggression, stylized ways of releasing energy.

Jupiter Behaviours involving risk-taking, grand and impressive gestures, exhibitionism, absence of inhibition.

Saturn Behaviours involving restraint, self-control, down-to-earth practicality, habitual and self-imposed inhibitions.

Uranus Behaviours involving clarity, precision, eccentricity, emphasis of one's uniqueness and originality.

Neptune Behaviours involving closeness to others, sympathetic gestures, submergence of one's individuality.

Pluto Behaviours involving ruthless dedication to a task.

H7 chart

Planets on the Ascendant/Descendant axis
Planets on this axis in the H7 chart will tend to show the person's romantic view of himself in his interactions with other people and with the world in general. He imagines himself playing a certain kind of part, which he invests with emotional significance. At times when he is under the influence of this inspiration, his actions will tend to take on the characteristics of the planet as listed below. If there is an opposition between a planet on the Ascendant and a planet on the Descendant, then it *may* be that the rising planet represents his romantic view of his own behaviour, while the setting planet represents his romantic view of the behaviour that other people tend to adopt in their dealings with him.

Planets on the M.C./I.C. axis
Planets on this axis in the H7 chart will tend to represent the romantic goals towards which the person feels he is moving.

Emotionally he feels that it is his mission to create the characteristics of the planets as listed below, either within his own behaviour or in the world outside. If there is an opposition between a planet on the I.C. and a planet on the M.C., then it *may* be that the planet on the I.C. represents the person's romantic view of the past (his own past or the world's past, or the past of that section of society which he wants to change), while the planet on the M.C. represents his romantic view of the future.

Sun Romantic self-fulfilment. (Sun on an Angle in the H7 chart indicates that the person has a romantic view of himself and that his imaginings and inspirations are concerned with his own self-fulfilment rather than with external matters.)

Moon Easily moved to tears and laughter by events in the environment and by the joys and trials of other people: emotionally responsive.

Mercury Having inspired thoughts and theories, able to bewitch with words: the Wise Old Man, the Prophet.

Venus Beautiful, loving, lovable, seductive: the Heroine.

Mars Dynamic, all-conquering, sexually irresistible: the Hero.

Jupiter Free, carefree, grand, noble, magnanimous, superhuman.

Saturn Bowed down by burdens, faithful, loyal: bearing the sorrows of the world.

Uranus Thrilling, brilliant, wild, shocking, rebellious, a law unto oneself.

Neptune Mystical, saintly, selfless, devoted to the service of others.

Pluto Ruthless, cruel, irresistible, awe-inspiring (but the ruthlessness may be turned on oneself as well as on others).

H9 chart

Planets on the Ascendant/Descendant axis

Planets on this axis in the H9 chart will tend to represent the type of happiness and joy which the person finds in his own interactions with other people and with the world in general. He is happy about the way in which he uses the qualities of the planet to relate to the world, and so he will tend to use those qualities in such a way as to spread his happiness to those around him. If there is an opposition between a planet on the Ascendant and a planet on the Descendant, then it *may* be that the rising planet represents the person's own behaviour, while the setting planet represents the qualities which the person believes he is able to develop in other people through the operation of the rising planet.

Planets on the M.C./I.C. axis

Planets on this axis in the H9 chart will tend to represent the type of happiness and joy towards which the person feels that he is working in his journey through life. If there is a religious or spiritual orientation, they can be said to represent his spiritual goals. If there is an opposition between a planet on the M.C. and a planet on the I.C., then it *may* be that the planet on the M.C. represents his goal of happiness in the future, while the planet on the I.C. represents the type of capacity for joy and happiness which he feels he brings with him from the past.

Sun Joy in self-fulfilment: joy in presenting oneself as one really is.

Moon Joy in being receptive and responsive to other people and to the world.

Mercury Joy in rational thought and communication.

Venus Joy in one's beliefs and values: joy in being able to attract (and repel) people as one wishes.

Mars Joy in one's physicality and dynamic energy: joy in one's ability to make decisions and to influence other people's actions.

Jupiter Joy in freedom. (This one word seems to capture the whole essence of Jupiter in H9.)

Saturn Joy in acceptance of one's earthly fate: joy in having an allotted role to perform.

Uranus Joy in one's unique individuality.

Neptune Joy in being a member of the human race: joy in being able to transcend one's own boundaries.

Pluto Joy in one's single-minded consistency and dedication.

PART THREE: CASES

Part Three consists of five case-studies. The first three are of famous people; the last two are of 'ordinary' people considered as astrological clients. In each case we will look at the radical chart and at the fourth-, fifth-, seventh- and ninth-harmonic charts, in order to show how a picture can be built up from a series of harmonic charts. The house system used in the radical charts is Placidus.

CHAPTER 12
ELEONORA DUSE

Eleonora Duse was, after Sarah Bernhardt, the most famous tragic actress of the late nineteenth century. She is given pride of place in this series of case-studies, because her chart so perfectly illustrates the function of each of the harmonic charts. Gauquelin quotes the following three passages from biographical accounts of Eleonora Duse:

> Eleonora Duse loathed all ham-acting and lived a simple life uncomplicated by love-affairs, completely injecting her entire self, body and soul, into her acting whenever she was before an audience. Off stage, she fiercely safeguarded her privacy.
>
> Eleonora Duse was admired for her intelligence, her fine breeding, and her depth of character. She was quite unlike most famous actresses. Her art did not consist of fabricating an artificial and exciting reality, but of representing specifically feminine feelings in their truest, most complete, but also starkest form. Someone justly wrote that 'the exercise of her art was a calling for her'. Her life was mysterious and secluded.
>
> 'Happiness,' said Miss Duse, 'would be to close one's door and, sitting alone at a table in a small room, create life in the isolation of life.'[1]

These three passages give an extraordinarily vivid and complete impression of Duse's personality and behaviour, which we can bear in mind as we examine her harmonic charts.

Radical chart (p.148)

The dominant impressions which we gain from Duse's radical chart are:

1 Moon and Saturn rising in Leo: sensitive and perceptive but

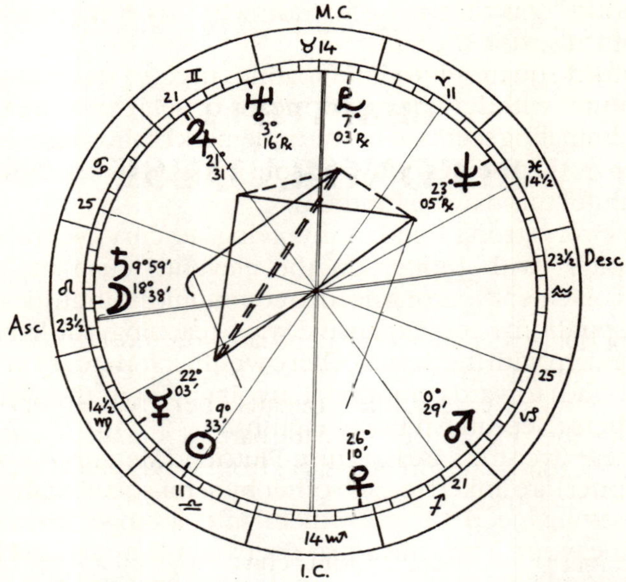

Eleonora Duse: Radical chart

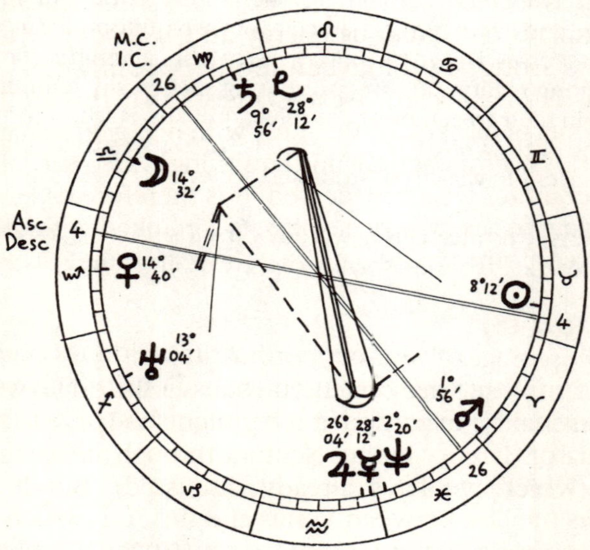

Eleonora Duse: H4 chart

proud, controlled and withdrawn: this is reinforced by Saturn sextile the Sun.

2 Saturn square Pluto culminating: urge to overcome her natural withdrawnness by means of self-exposure (Pluto culminating seems to be the trademark of those artists who throw themselves body and soul into their work, however painful this may be for them).

3 The very strong T-square involving Mercury, Neptune and Jupiter (with Jupiter on the eleventh cusp of 'social involvement'): strong need to communicate in an idealistic, inspirational and expansive way, reaching out from her concern with the details of her own affairs (Mercury in Virgo second) towards emotional involvement in the affairs of others (Neptune in Pisces eighth).

4 These two stress areas (Saturn-Pluto and Mercury-Neptune-Jupiter) are linked to each other by semi-squares, showing a strong need to unite them into a single pattern of endeavour. The semi-square between Mercury and Pluto is completely exact (and so will reappear as an exact aspect in all the harmonic charts): this urge for self-exposure through self-punishing communication is the single most compulsive drive in Duse's personality.

5 Mars unaspected: according to Dean[2] this should mean that she displays conspicuous energy and 'a tendency to be always starting new things before others are finished'. Also Uranus is unaspected: Dean suggests that this leads to 'moments of marked restlessness with no apparent cause, often a love of innovations and new things'. Also Sun, Moon and Venus are weakly aspected. It is therefore especially important to see how these planets are linked with others in the harmonic charts.

H4 chart (p.148)

This chart throws into sharper relief the stresses and drives within Duse's personality. In particular it highlights the urge to fulfil the potential of Mercury-Jupiter-Neptune through the operation of Pluto (which we have already discussed). But it also emphasizes the link between Venus and the Sun (which was present in the radical chart as a semi-square): in the H4 chart we can see Venus conjunct the Ascendant (striving to express her femininity), but we also see Venus and the Ascendant opposite the Sun, showing that through this expression of her femininity

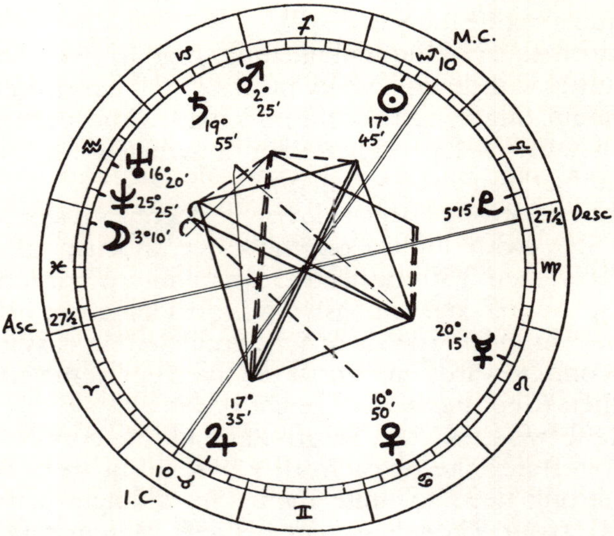

Eleonora Duse: H5 chart

she strives to integrate her personality and find her true self.

The Moon has a number of aspects in the H4 chart: in particular, the semi-squares to Mercury and Pluto show that at a deep level Duse is aware of the need to utilize her sensitivity and receptivity (Moon rising in the radical chart) in her compulsive drive towards ruthless communication (Mercury-Pluto). But Mars, which was unaspected in the radical chart, is still almost entirely unaspected in the H4 chart (although its opposition to the M.C. shows that Duse will strive to express Martian energy in her work), so we must look to other harmonics to find how Mars is integrated with other planets.

H5 chart (p.150)

Duse's H5 chart is extremely strong and well-integrated. This is what we expect (and usually find) in the charts of famous actors and actresses. Even though Gauquelin's quotations say that Duse 'loathed all ham-acting' and that 'her art did not consist of fabricating an artificial and exciting reality', it is clear from the quotations that Duse *did* create for herself a different personality when she was on stage, a personality in which her normal solitariness and withdrawn-ness were set aside, and in which she was able to 'inject her entire self, body and soul, into her acting'. The H5 chart seems to represent her on-stage personality.

The links in Duse's H5 chart are mainly stressful ones, showing how Duse's on-stage personality was one of stress and conflict (the tragic as opposed to the comic actress). The stressful links from Mercury to Jupiter, Neptune and Pluto are carried over from the radical chart. But, in the H5 chart, Mercury and Jupiter are linked with Uranus and the Sun to form a Grand Cross, with an almost exact opposition between Jupiter and the Sun, from the I.C. to the M.C. Thus the urge for expansive communication is linked with the urge to discover and project her unique individuality. We can see in this Grand Cross the desire for fame, the need to 'see one's name in lights', which must have driven Duse out of her isolation on to the stage, and kept her there despite her desire to retreat into her privacy (for the fifth harmonic is always to do with habit-formation, the formation of compulsive patterns). But also we should note the Sun-Mercury square, which is indicative of an obsessive drive towards self-understanding, and suggests that Duse went on to the stage partly in order to know herself more fully. I believe it was the involvement of the Sun in this Grand Cross which ensured that Duse's acting was 'for real' and was not a 'fabrication of an artificial reality': she needed to reveal her true nature through her stylized behaviour.

The character of Uranus in the Grand Cross is affected by its conjunction with Neptune and the Moon. Here we see something of the 'reality' which Duse was trying to convey: the representation of 'specifically feminine feelings in their truest, most complete, but also starkest form'. (Neptune is also linked by semi-square with Venus.)

But also, Mars is square Pluto, and Mars and Pluto are linked by semi-squares with the planets in the Grand Cross: the exact semi-square from Mercury to Pluto (carried over from the radical chart), and the semi-squares from Mars to Jupiter (exact), the Sun and Uranus. Mars is also sextile the Moon: so that here in the H5 chart we see at last how Duse succeeded in integrating her Martian energy into the rest of her personality. In order to do so, she needed to create a fifth-harmonic pattern of behaviour: a pattern deliberately formed out of the creation of habits and structures. The theatre provided Duse with the means of doing this. Within her 'stage personality', she was able, through the outpouring of Martian energy, to find complete fulfilment for the whole of her nature: outside her 'stage personality', she

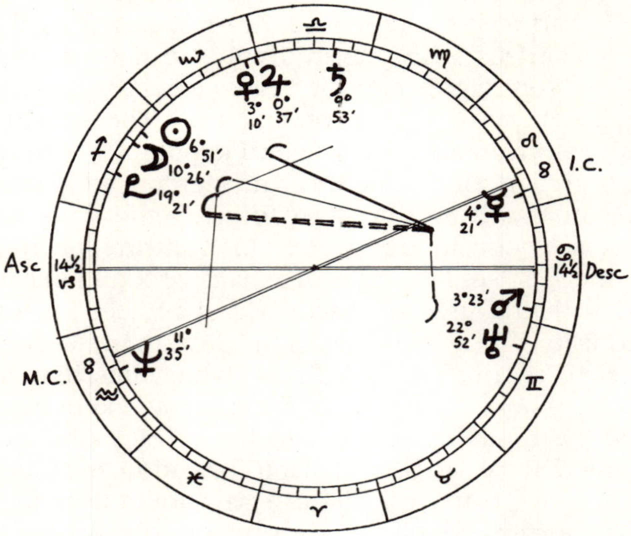

Eleonora Duse: H7 chart

tended to show a lack of Martian vigour. It is interesting to speculate on how Duse would have chosen to realize the potentialities of her H5 chart if theatrical opportunities had not been open to her. It is possible that she would have tried to fulfil herself through sport, which offers rules and patterns for the expression of Martian energy: but sport would have offered far less fulfilment for the Moon-Uranus-Neptune side of the Grand Cross, and she would probably not have been so successful in this field.

The only planet in the H5 chart which has 'easy' rather than 'difficult' links with other planets is Saturn, which is trine to Jupiter and widely sextile to the Sun. This suggests that within Duse's 'stage personality' the difficulties and strains associated with Saturn (Saturn square Pluto in the radical chart; Saturn square Uranus in H4) evaporated, and Duse was able to use her natural coolness and withdrawn-ness easily and harmoniously for the purpose of tempering, controlling and directing the extravagances of her Sun-Jupiter drives. Saturn helped her to avoid 'ham-acting' and to introduce the necessary restraint and self-discipline into her acting. But the quincunx between Saturn and Mercury suggests that, within her acting style, there was an unresolved conflict between this restraint and the obsessive desire for communication.

The final point to be made about Duse's H5 chart is that Pluto is close to the H5 Descendant. Any planet on the H5 Ascendant or Descendant seems to say something about the way in which the person portrays himself to the world. Duse portrays herself as a Plutonian person: a person who penetrates to the depths of whatever she is concerned with.

H7 chart (p.152)

Duse's H7 chart is also very strong: not so strong as the H5 chart, but still containing a number of very striking features. The seventh harmonic, like the fifth, is necessary for the understanding of Duse's acting talent. Whereas the H5 chart says most about her acting *style,* the H7 chart says more about the *inspiration* on which she drew and hence about the *content* of her acting.

Perhaps the most noticeable feature is the Sun-Moon-Pluto conjunction. The harmonic in which the Sun and the Moon come together is the one in which the person achieves the greatest integration of his personality, and hence it says something about the person's vision of himself. Duse's Sun-Moon conjunction in H7 shows that she has a romantic vision of herself: within the romantic mode she achieves the self-sufficiency, the state of being at peace with herself, which is typical of a Sun-Moon conjunction. But the conjunction of Sun and Moon with Pluto shows that this is a vision of herself exposed, herself overturned, herself relentlessly driven forward towards re-birth and renewal. So this will drive her further towards total self-absorption in her acting and towards the ruthless exposure of her deepest feelings. The emphasis on Pluto, which we have noted throughout this analysis (Pluto on the radical Midheaven, Pluto exactly semi-square Mercury, Pluto on the H5 Descendant) is thus further reinforced.

And both Pluto and the Sun are linked with Mercury in the H7 chart: Pluto by the original semi-square, and the Sun by a trine which makes it easy for Duse to communicate the nature of her true self in a romantic and uninhibited way. (We should note that Mercury is both square the Sun in H5 and trine the Sun in H7: that is to say, it is conjunct the Sun in both the twentieth and the twenty-first harmonics. (It is likely that this combination – which will occur whenever the Sun and Mercury are seventeen and a half degrees apart in the radical chart – has a special significance, since it gives the person both the obsessive drive

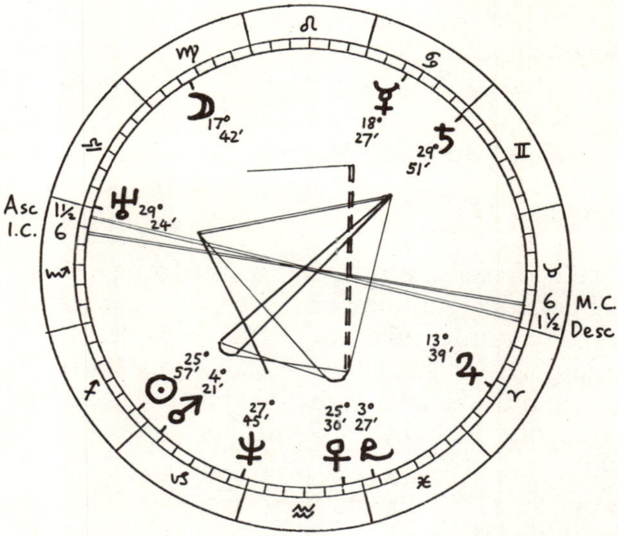

Eleonora Duse: H9 chart

towards structured self-understanding and the ability to communicate this self-understanding in a free and unstructured way.)

But Mercury is on the H7 I.C., widely opposite to Neptune on the H7 M.C.: Duse sees herself (within her romantic vision) as a naturally rational and communicative person whose aim is to rise out of this rationality into idealism and transcendence. Neptune on the M.C. in the H7 chart seems to show that Duse wants to achieve mystical ends, and to be seen as a person of mystery: but the nature of this idealism and transcendence is illuminated by the sextile and trine from Neptune to the Moon and Saturn. It is an idealism involving self-discipline and control of feelings. As the Gauquelin quotations say: 'she lived a simple life uncomplicated by love-affairs . . . she fiercely safeguarded her privacy . . . her life was mysterious and secluded'.

We should also note the importance of Venus in Duse's H7 chart. Venus is conjunct Jupiter, and Venus and Jupiter are trine to Mars, with Venus also square Mercury: the expansion of femininity, the determination to communicate its nature, and the ease in expressing it through vigorous action. 'Her art consisted in representing specifically feminine feelings in their truest, most complete, but also starkest form.'

H9 chart (p.154)

In the H9 chart we see an entirely different pattern. Mercury was strong in all the previous harmonics ('she was admired for her intelligence'); but here, despite the persistence of the Mercury-Pluto semi-square and despite the Mercury-Moon sextile, Mercury stands clearly apart from the main configuration of planets; Moon is also weakly aspected, and Jupiter is entirely unaspected. All the emphasis is on the Grand Trine involving Uranus (on both the Ascendant and the I.C.), Venus, Pluto and Saturn, with Saturn (the pivotal planet) opposite the Sun and Mars.

This seems to show that for Duse, happiness and joy (which are the province of the ninth harmonic) mean the casting-off of rationality (Mercury), of expansion and growth (Jupiter), and of responsiveness to other people (Moon), and the entry into a world in which she stands alone as a separate and unique individual (Uranus), rejoicing in her limitations and restrictions (Saturn), and rejoicing also in the complete exposure of her own femininity (Venus-Pluto). Yet the opposition between Saturn and Sun-Mars seems to show that, by restraining and enclosing herself, Duse is able to rejoice in her own power. At a rational level the meaning of this statement is scarcely clear; but at a more mystical level it seems to have meaning, and this meaning is perfectly expressed by the Gauquelin quotation: 'Happiness,' said Miss Duse, 'would be to close one's door and, sitting alone at table in a small room, create life in the isolation of life.'

CHAPTER 13
HENRI ROUSSEAU

Henri Rousseau (otherwise known as le Douanier Rousseau) was the greatest of the modern 'primitive' or 'naive' painters, and the first of them to be well-known. He did not start painting seriously until after the age of forty, and he was almost entirely self-taught. His painting technique was in many ways restricted: for instance, he always had difficulty with the laws of perspective. But he succeeded, by hard work and elaborate attention to detail, in creating a style that was uniquely his own.

The quality which shines from Rousseau's best paintings, and gives them their extraordinary power, is a kind of magical and dream-like stillness. Ronald Alley says:

> In Rousseau's later jungle pictures there is a frozen stillness. The vegetation is profuse and luxuriant, yet everything is beautifully ordered and lucid. Each form is depicted with sharp definition, almost every leaf turned towards us to show its characteristic shape. Yet in spite of the profusion of details, each picture is a perfect unity: nothing jumps, everything is in its place. There is always something going on in the jungle, usually either some amusing monkeys at play or an incident of brutal savagery, such as a negro attacked by a jaguar or a tiger attacking a buffalo. The mood that the pictures evoke, with their strangeness and latent menace, is extraordinarily convincing, yet the origin of Rousseau's inspiration remains an enigma.[1]

In fact it seems that Rousseau obtained the models for his tropical animals and plants partly from the Paris zoo and partly from an animal book for children. But the real inspiration comes from within: the paintings are a record of Rousseau's inner world, and of his inner feelings about the outer world. Alley says: 'According to Apollinaire, Rousseau was sometimes so overcome when painting works like these that he would become terrified and, all-a-tremble, would be obliged to open the window'.[2]

In addition to his paintings, Rousseau played the violin, and he also wrote a play called *Revenge of a Russian Orphan Girl* in which there is a stark contrast between good and evil characters, and good triumphs over evil.

Outside his paintings, Rousseau led a dull and frustrating life. For most of his adult life he performed (not very well) a routine clerical job. He had six children, five of whom died young. After the death of his second wife, he lived in great squalor, without water or lighting, and with a tattered carpet on the floor which a shopkeeper had given him in exchange for three of his paintings. Towards the end of his life he received some recognition (of a very patronizing kind) from artists such as Picasso, but he remained in severe poverty until his death at the age of sixty-six.

Alley says: 'The innocence of Rousseau, his unworldliness and his childlike trustfulness, is attested by all those who knew him, and there are many stories of the tricks played on him by people who could not resist the temptation to tease and mock him'.[3] Rousseau in return never hurt anyone, but he sometimes resorted to 'little lies' in order to protect his pride. He was prone to inventing stories about his own exploits as a young man, and he seems himself to have half-believed these stories. He was twice imprisoned, once for petty theft and once for involvement in a bank fraud: it seems that these misdemeanours can be put down to his naivety and his ignorance of the ways of the world.

Rousseau was carried forward above all by a belief in his own genius. He said to Picasso: 'We are the two great painters of the age'[4]; and he wrote, 'If my parents had known that I had a gift for painting, I should now have been the greatest and richest painter in France'.[5]

Radical chart (p.158)

Rousseau's radical chart explains many points about his nature, which we will discuss very briefly. Moon conjunct Venus in the fifth house says much about his desire to create beautiful things which reflect his own emotional state. Neptune rising, with Saturn in the twelfth house, speaks of his unworldly manner and his feelings of inferiority (and also superiority) in relation to others.

But, from the point of view of harmonic analysis, the most interesting feature is the configuration of Mars trine Neptune, with both Mars and Neptune sextile to Pluto. These three planets

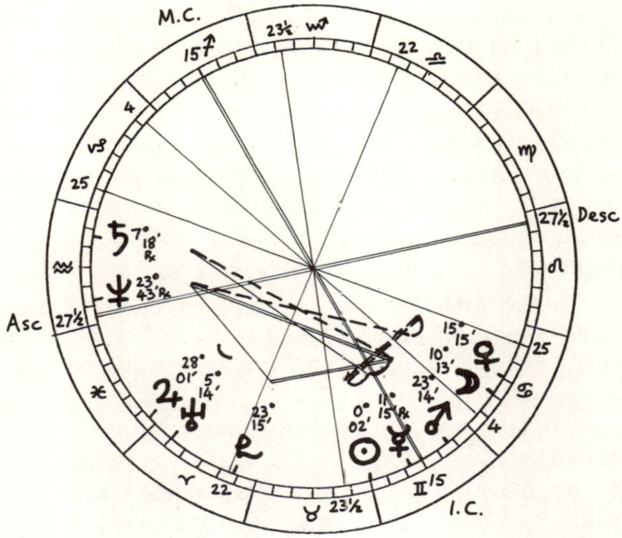

Henri Rousseau: Radical chart

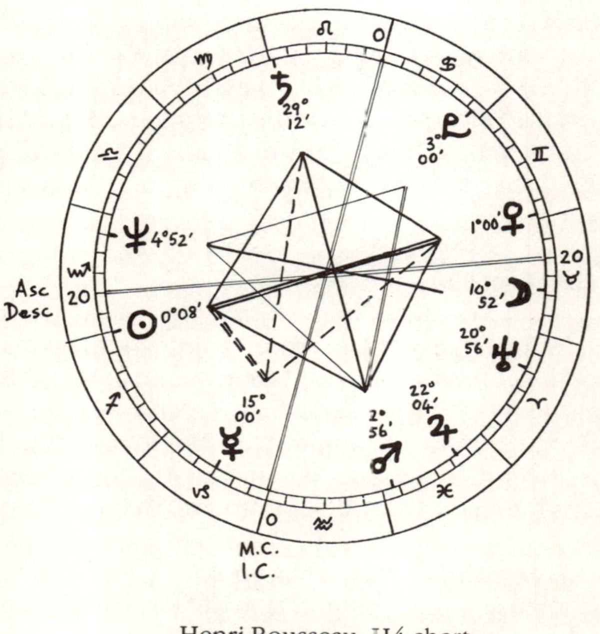

Henri Rousseau: H4 chart

are linked by very close aspects. In particular, the Mars-Pluto sextile is virtually exact and so can be seen as the most important aspect in the chart, since it will reappear as a close aspect in all the harmonics. Mars sextile Pluto can be literally interpreted as 'seeking pleasure through ruthless (self-punishing) expenditure of energy'; and the involvement of Neptune means that this effort is connected with the need to transcend his own individuality and to achieve imaginative meaning. Painting is a physical activity, and Mars is often the indicator of the painting task in the charts of painters. Rousseau himself described the process by which his paintings were produced as one of 'unremitting toil'; but the fact that these planets are linked by trines and sextiles shows that this labour was a labour of love and of pleasure.

There are also the three semi-squares (Sun-Venus, Moon-Neptune and Mars-Saturn), which are more clearly shown in the H4 chart and so will be discussed under that heading.

The two least-integrated planets in the radical chart are Jupiter and Uranus, and it is necessary to look at the harmonic charts to see how Rousseau integrated these planetary forces into his behaviour.

H4 chart (p.158)

The H4 chart contains a Grand Cross involving the Sun. This immediately suggests that Rousseau had severe problems in relating to the world, which, if unresolved, could possibly lead to mental illness. On the one hand, there is the close opposition between Sun and Venus, showing that Rousseau had a strong need to love and be loved, to appreciate and be appreciated, based on an underlying fear that he was unlovable. (The strength of this need is shown by Rousseau's great desire for recognition as a painter, and also by the fact that even in his sixties he was writing passionate letters to a lady-friend, full of the sorrows of unrequited love.) But at right-angles to this opposition is the opposition between Mars and Saturn. This second opposition in itself suggests only that Rousseau was striving to impose Saturnian control and discipline on his Martian activities, including his painting activities. But the fact that this opposition cuts across the Sun-Venus opposition suggests that it acts as a kind of block or barrier to the fulfilment of Rousseau's need for love and appreciation. Since problems in H4 are often externalized (projected on to other people), I would suggest that the Mars-Saturn opposition represents Rousseau's vision of the

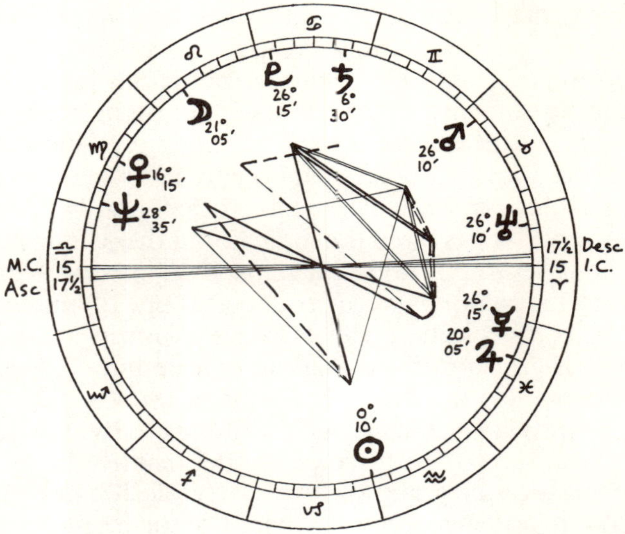

Henri Rousseau: H5 chart

harshness and cruelty of the outside world, and also helps to explain the fact that by his behaviour he seemed to invite a cruel and mocking response from other people. Rousseau's life presents a stark contrast between his own loving nature (Sun-Venus) and the harshnesses and restrictions imposed on him from outside (Mars-Saturn); and the same stark contrast between good and evil is found in his works (on the one hand, the happy monkeys playing in the forest; on the other hand, the tiger ripping the flesh of the buffalo, or the goddess of War riding victorious over the dead bodies while the crows tear at their flesh).

Also in the H4 chart, we can see that Mercury is linked by semi-squares to three of the planets in the Grand Cross, showing that Rousseau is constantly striving to *make sense* of this whole complex of problems and to communicate about it to others. Without this Mercury-activity, Rousseau might have been unable to find an outlet for his fears in his paintings, and might have developed severely neurotic tendencies (though this is offset by the Sun-Moon conjunction in H9, which we will come to later on.)

The configuration of Mars, Neptune and Pluto (which we noted in the radical chart) appears in the H4 chart as a Grand

Trine, and is linked (through Neptune) to the Moon, showing that Rousseau is trying to express his feelings in ways that have meaning for the whole of humanity.

Jupiter and Uranus are still unaspected in the H4 chart, showing that at this level Rousseau is unable to integrate the search for growth (Jupiter) and the search for uniqueness (Uranus) into the rest of his behaviour. This may well be an indication of the humdrumness of Rousseau's life outside his painting. In order to understand how Rousseau seeks growth and uniqueness, we have to turn to the later harmonics which are more closely connected with creative talent.

H5 chart (p.160)

The most striking feature of Rousseau's H5 chart can be easily seen. The exact Mars-Pluto sextile (carried over from the radical chart) is linked by *exact* aspects with both Mercury and Uranus (all four of these planets are at 26° 10′ or 26° 15′ of a sign). This union of four planets by *exact* aspects is most unusual, and is the most remarkable feature in any of Rousseau's harmonic charts. It shows that Rousseau was able to develop (and *enjoyed* developing – Mercury trine Pluto) a style in which the characteristics of these four planets were completely united: Mars and Pluto for 'unremitting toil', Mercury for careful thinking-out, and Uranus for clarity, brilliance and precision. (Uranus is especially important because it is the only planet close to both the Ascendant-Descendant and the M.C.-I.C. axes.) This exactly describes Rousseau's painting style (the *process* by which he created his paintings).

We can note also that Mars is trine to Neptune and to the Sun (Sun being also opposite Pluto), showing the great extent to which Rousseau was able to find self-fulfilment through the development of his painting style (Mars being, as we have said, the significator of the painting act itself).

Mercury is also conjunct Jupiter, and Mercury and Jupiter are opposite Venus and Neptune. We can express this as follows: on the one hand, a style expressing an abundance of ideas; on the other hand, a style expressing universal beauty. The tension between the two explains much of the vitality of Rousseau's paintings.

Moon and Saturn are weakly integrated into the H5 chart. In the case of the Moon, this indicates Rousseau's difficulties in integrating his *perceptions* into his style (hence, perhaps, some

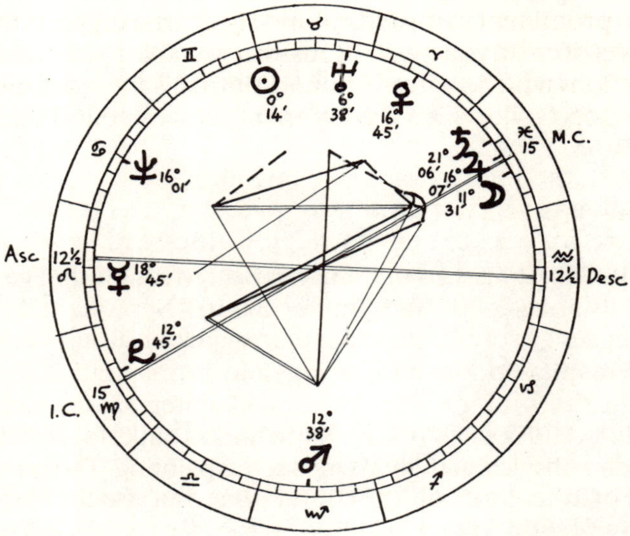

Henri Rousseau: H7 chart

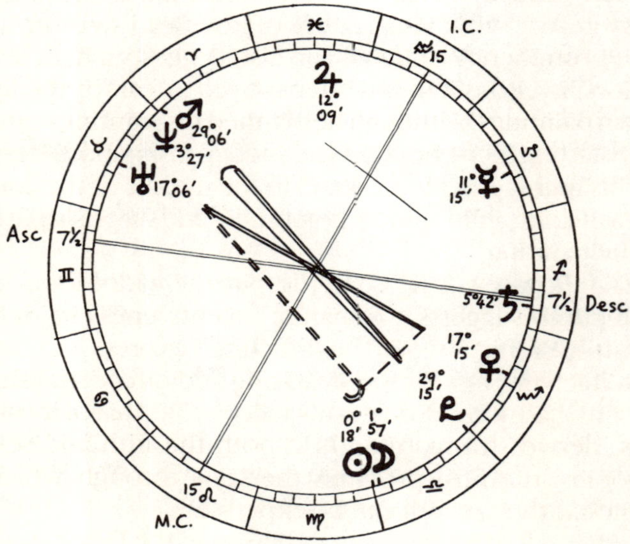

Henri Rousseau: H9 chart

of his limitations as a painter). But the isolation of Saturn (which was so prominent in the radical and H4 charts) suggests that, at the level of creative activity, Rousseau was able to cast aside the restrictions which so greatly bound him in his everyday life, and to develop a style which was luxuriant, expansive and free from inhibition.

H7 chart (p.162)

The Moon in Rousseau's H7 chart is near to the Midheaven and is closely linked with Mars, Jupiter, Saturn, Neptune and Pluto. This suggests a very fertile romantic imagination: Rousseau is able to link his perceptions and feelings (Moon) inspirationally (H7) with the four great moral principles of Jupiter, Saturn, Neptune and Pluto, and he is also easily able to find an outlet for his feelings through a physical activity (Mars) such as painting. The predominance of trines (not only the Grand Trine, but also the Mercury-Venus trine suggesting 'inspired charm of expression') shows that Rousseau's fantasy-world is in the main a happy and joyous one; but the opposition between Pluto and Moon-Jupiter on the I.C.-M.C. axis shows Rousseau's need to introduce brutality and savagery into his happy jungle scenes; and the close square between Venus and Neptune is an indication of the tense and uneasy beauty which is a feature of Rousseau's vision.

The prominence of the Moon also suggests a high level of emotionality: Rousseau was often moved to tears by the sight of his own paintings. Although he planned his paintings carefully before starting them, he often had sudden inspirations (resulting from his highly emotional state) in the middle of the painting process, leading him to incorporate new and surprising features into the painting.

In fact, Rousseau's H7 chart (despite the wide Moon-Saturn conjunction) suggests a romantic inventiveness so rich and uncontrolled that it would be difficult to express it in creative terms, if it were not for the steadying and directing influence of the fifth harmonic. Rousseau's success, like that of most great artists, derives from strength in both the fifth and seventh harmonics: the fifth provides the style through which the fantasies of the seventh can be expressed.

But the lack of integration of the Sun in the H7 chart shows that, in a sense, Rousseau himself is not involved in his own imaginative world: his fantasies are not about himself, but about his interpretation of the world he sees around him.

H9 chart (p.162)

By far the most important feature of Rousseau's H9 chart (and perhaps of any of his harmonic charts) is the close Sun-Moon conjunction, which shows that Rousseau was – and was seen to be – an essentially ninth-harmonic person: a person of joy and happiness, purity and innocence, a spiritual rather than a worldly person. This was the first quality that impressed all who met Rousseau, and it shines vividly through his paintings, in spite of all the complexities of his style and his fantasies.

Yet Sun and Moon are not well integrated with the other planets in the H9 chart, suggesting that, despite his instinctive spirituality, Rousseau was not able to work out a consistent spiritual philosophy. His instinctive nature was spiritual, but his imagination and his conscious thoughts focussed on worldly things. He was instinctively happy, but consciously unhappy.

The Mars-Neptune-Pluto opposition is carried over from the radical chart. The other striking feature of Rousseau's H9 chart is the close opposition between Venus and Uranus, which may well be an indicator of Rousseau's tendency, even in old age, to fall desperately in love with younger women and to express his tenderness towards them in pure and naive language. But the fact that this opposition occurs in H9 (rather than in H7) suggests that this was not the result of Rousseau's fantasies, but rather the result of his genuine perception in these women of a quality of 'pure femininity' which he needed to complement his own individuality.

CHAPTER 14
LUDWIG VAN BEETHOVEN

In the first draft of this book, Ludwig van Beethoven played an important role. I presented his fourth-, fifth-, seventh- and ninth-harmonic charts in Chapters 4, 5, 6 and 7 and used them as the connecting links between these chapters. I was therefore distressed to learn from *Transit*[1] that there is in fact no record of Beethoven's time of birth, and that all published charts must therefore be regarded as speculative. In view of this I clearly could not give Beethoven the central role which he had had in the first draft.

However, it is of course possible that there was some evidence of the birth time which has now been lost. The birth time which is most frequently quoted (1.30 p.m. on 16 December 1770) has been used by astrologers at least since 1934, and attempts to discover the original source of the information have so far proved unsuccessful. It may have been originally a speculative chart, but it is also possible that it was based on biographical evidence. To me, and to many other astrologers, it seems an extremely convincing chart, and the harmonic charts are very illuminating. I have therefore decided to retain a brief account of Beethoven's harmonic charts in this chapter: but I must ask readers to treat these charts with caution, because of the uncertainty of the birth time.

Radical chart (p.166)

I will not analyse Beethoven's radical chart in detail, but will simply point to its most outstanding feature, which is the close opposition between Mercury and Mars, with Mercury enclosed by the Sun and Moon. The Mercury-Mars opposition (which is so close that it recurs in all the harmonic charts), in the restless signs of Sagittarius and Gemini, shows a constant searching for forceful communication and for communicative action which is the main driving-force in Beethoven's life; and the

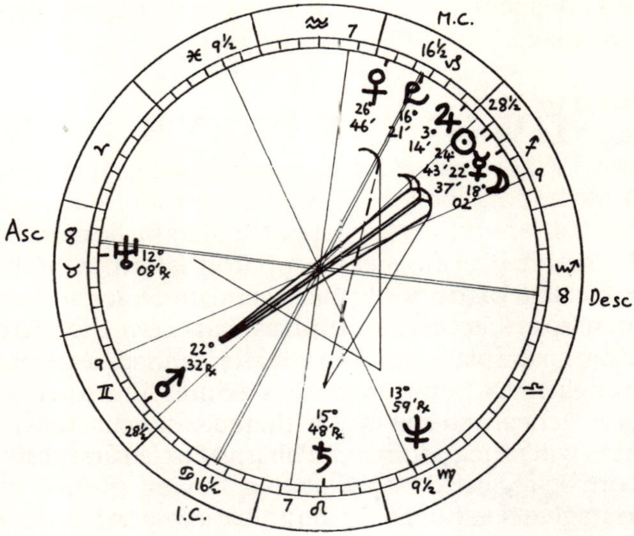

Ludwig van Beethoven: Radical chart

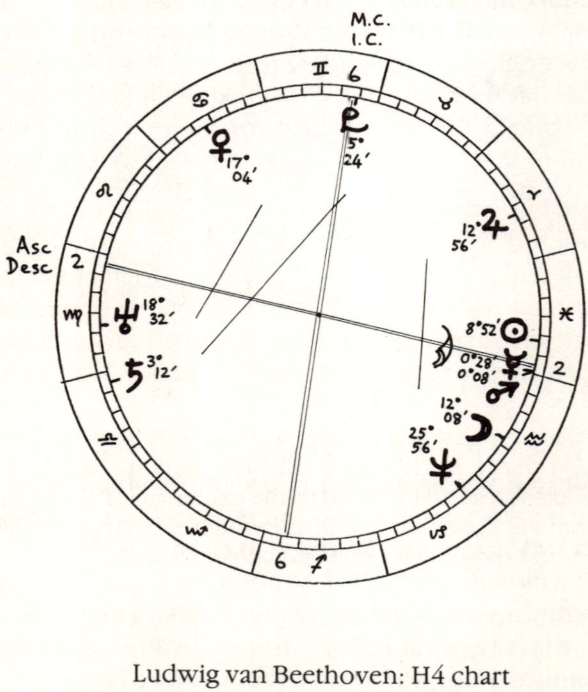

Ludwig van Beethoven: H4 chart

conjunctions of Mercury with Sun and Moon show Beethoven's total involvement in this search, to the extent that it takes over his whole life.

H4 chart (p.166)

Beethoven's H4 chart seems surprisingly weak. Its only strong feature is the conjunction of Sun, Mercury and Mars, which is the result of Sun and Mercury opposite Mars in the radical chart; and the importance of this 'cluster of effort' is further emphasized by the fact that it is opposite the Ascendant/Descendant point. But apart from this, there are no 'hard' aspects in the chart at all, and only three isolated and not very strong 'soft' aspects. If we allow slightly wider orbs, there is a Sun-Pluto square and a Mercury-Venus semi-square, but there are no other aspects that come anywhere near inclusion. This is, in fact, an unusually weak H4 chart.

This might seem to destroy the whole of our theory about the fourth harmonic. Beethoven was after all an outstanding example of a 'striver'. In Scholes's words:

> Beethoven composed slowly and with much erasure . . . . The impression . . . is that of a powerful individuality with difficulty finding its adequate means of expression, and at last, after efforts that possibly surpass those of any other composer who ever lived, forcing its way into the light of day. [2]

In order to make sense of Beethoven's weak H4 chart, we have to compare him with the people with strong H4 charts whom we considered in Chapter 4. All of them were striving to accomplish things in the world, in relation to their environment and in relation to other people. This was not true of Beethoven. Scholes presents a picture of Beethoven sitting at his desk in his later years, surrounded by indescribable squalor. He comments:

> The squalid disorder meant nothing to him in those days. He had finished with the world. Since 1824 the medium of the string quartet had absorbed his mind to the exclusion of all else, and now, stone-deaf, very ill but still indomitable, he rose to heights which even he had never reached before. [3]

'He had finished with the world'. One cannot imagine people with strong H4 charts, like Eddy Merckx or Winston Churchill, finishing with the world. Beethoven's struggles were of a quite

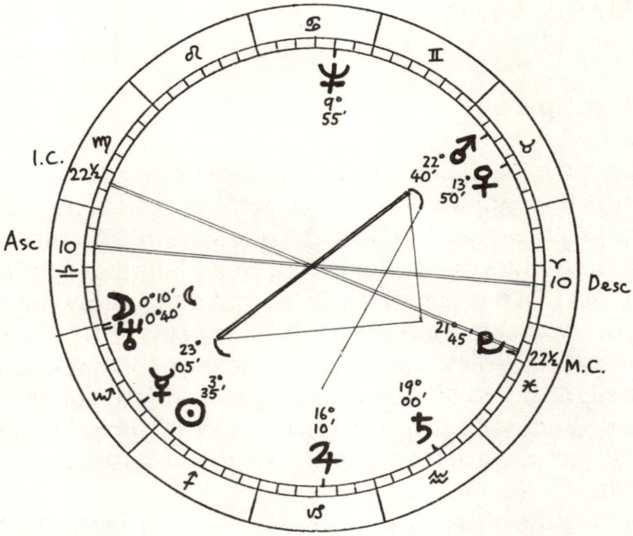

Ludwig van Beethoven: H5 chart

different nature. In his dealings with the world, Beethoven was careless and confused. He was not concerned with identifying 'enemies', he would probably not have troubled to fight them. J.W.N. Sullivan says:

> Among the elements constitutive of Beethoven's personality we must include his lack of malleability. This quality made him almost immune from purely external influences. Thus he was impervious to criticism; his manners were atrocious; he ignored conventions; he was permanently subject to no social passions, not even sexual love . . . . He was not an educable man.[4]

A man who was strong in the fourth harmonic could never be 'immune from external influences' as Beethoven was, for he would see himself as constantly struggling with those influences. But Beethoven's struggles were with his own inner creative inspiration. In order to understand this astrologically we must turn to his H7 chart, which is very strong and contains a large number of 'hard' aspects. Beethoven's harmonic number is not 4, but 28 (7 × 4).

H5 chart (p.168)

Beethoven's H5 chart is almost as weakly integrated as his H4 chart. The Mercury-Mars opposition is carried over from the radical chart, so that the only 'new' elements in the H5 chart are: firstly, the 'easy' links between Pluto and Mercury-Mars; secondly, the Venus-Jupiter trine (and the wide Venus-Mars conjunction); and finally the close Moon-Uranus conjunction, which, although it is very powerful in itself (giving Beethoven a great receptivity to original and striking musical forms), is disconnected from the rest of the chart. Beethoven, unlike Mozart (whose strong H5 chart we presented in Chapter 5), was not a natural stylist, and the creation of form and order did not come easily to him. Outside music, the extreme untidiness of his surroundings was notorious. Within music, he composed very slowly and with constant revisions and erasures; he did not have within his mind the free flow of rhythms and melodies which occurred naturally to Mozart, and the reduction of his musical inspiration to an ordered composition was always difficult for him. Beethoven's struggles were the expression of the close Mercury-Mars opposition in his radical chart (the struggle to communicate energy), but the only help given to him by his H5 chart was the link between Pluto and Mercury-Mars, which meant that he was easily able to adopt a style of relentless application to the task.

H7 chart (p.170)

It is in the seventh harmonic that Beethoven at last comes into his own. Beethoven's H7 chart presents an awesome spectacle of conflict, commensurate with the grandeur of his music. All ten planets are locked together in a complex pattern of aspects, yet there are no conjunctions: Beethoven's planets do not come together in the seventh harmonic, but it is in this harmonic that they become aware of each other and are able to act out the drama between them.

A whole chapter could be devoted to the unravelling of the complexities of this chart, and I will not attempt to do so here. But a few points stand out. Especially notable is the close involvement of all five of the outer planets: Jupiter, Uranus and Pluto through a tight T-square (this type of conflict relationship between the outer planets in H7 appears to be typical of the great composers), and also through Sun trine Uranus and Sun sextile

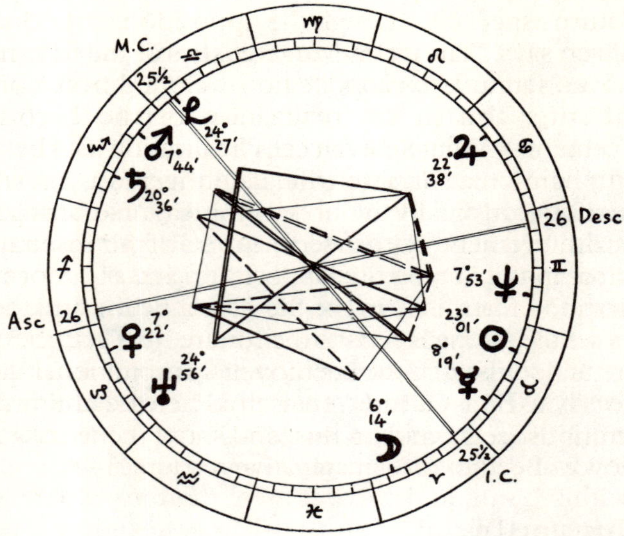

Ludwig van Beethoven: H7 chart

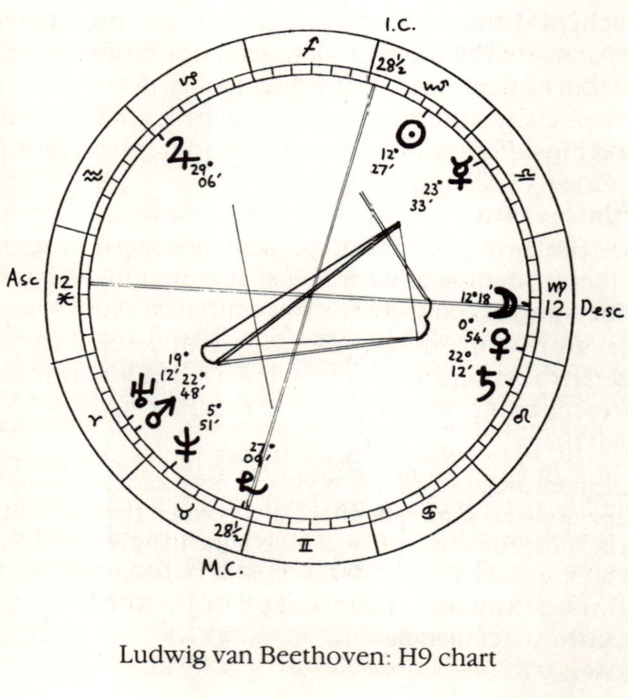

Ludwig van Beethoven: H9 chart

Jupiter; Neptune through a complex set of links with five planets; and Saturn especially through the opposition to the Sun.

Sullivan says: 'The chief characteristics of the fully mature Beethoven's attitude towards life are to be found in his realization of suffering and in his realization of the heroism of achievement'.[5] What Sullivan calls 'realization' is, I believe, a seventh-harmonic characteristic: Beethoven was *inspired,* or *aroused* to creative endeavour, by the ideas of suffering and of heroic achievement. We can see the roots of the 'inspiration of suffering' in the Sun-Saturn opposition, and the roots of the 'inspiration of heroism' in the Sun-Uranus trine and the Sun-Jupiter sextile. These aspects to the Sun are, I believe, the central feature, since they show Beethoven's inspirational view of himself (himself as sufferer, himself as heroic achiever). The remaining aspects build on this, and show the extraordinary richness of Beethoven's imaginative resources.

H9 chart (p.170)

Beethoven's greatest strength, as we have seen, was in the seventh harmonic. And yet we have said that the ninth harmonic is connected with joy; and joy is a quality which is conveyed by much of Beethoven's music – not only the Choral Movement of the *Ninth Symphony,* in which he made use of Schiller's 'Ode to Joy', but in many others of his later works. Beethoven, though he is inspired by suffering, seems to have the ability to reach out beyond the suffering into a state of pure joy and serenity. In fact, according to Sullivan:

> The opening Fugue (of the C sharp minor string quartet) is the most superhuman piece of music that Beethoven has ever written. It is the completely unfaltering rendering into music of what we can only call the mystic vision. It has that serenity which, as Wagner said, speaking of these quartets, passes beyond beauty. Nowhere else in music are we made so aware, as here, of a state of consciousness surpassing our own, where our problems do not exist, and to which even our highest aspirations, those that we can formulate, provide no key. Those faint and troubling intimations we sometimes have of a vision different from and yet including our own, of a way of apprehending life, passionless, perfect, and complete, that resolves all our discords, are here presented with the reality they had glimpsed . . . . That passionless, remote calm can seem, as it did to Wagner, like a melancholy too profound for any tears. To Berlioz it was

terrifying. To Beethoven himself it was the justification of, and the key to, life. In the light of this vision he surveys the world.[6]

What can we see in Beethoven's H9 chart that would help to explain this ability to rise to such a state of consciousness and express it in music? At first sight Beethoven's H9 chart may seem disappointing. It does not contain as many aspects as the strong H9 charts which we looked at in Chapter 7. And yet it has a clear and strong structure. The Mercury-Mars opposition (carried over from the radical chart) is, as we have said, the central clue to Beethoven's whole personality; but in the H9 chart Uranus is conjunct Mars and opposite Mercury, which shows that Beethoven can communicate energy clearly and vividly in the ninth-harmonic mode; and Saturn (the same Saturn which in the H7 chart signified suffering) is trine Mars-Uranus and sextile Mercury, showing that Beethoven can transmute suffering into serenity, and experience joy in so doing.

But the strongest clue is the almost exact sextile between Sun and Moon. Sun-Moon contacts are especially important, and the harmonic in which the Sun and the Moon come together seems, in a sense, to be the harmonic to which a person naturally belongs, even though he may be more *active* in one of the other harmonics. Beethoven's Sun and Moon have not come together in any of his previous harmonic charts, but they come together in an exact sextile in the H9 chart, which would appear as an exact opposition in the H27 chart. Twenty-seven is $3 \times 3 \times 3$. Thus, conjunctions in H27 (which are trines in H9) signify an even purer type of joy, which can perhaps be called 'ecstasy' or 'serenity', though there may be no word to describe it adequately. Beethoven, with his exact Sun-Moon sextile in H9, is striving towards this type of joy.

Nevertheless, there may be many people with Sun-Moon sextiles in H9 who never experience such an exalted state of consciousness (still less communicate it through art). The main creative force in Beethoven's personality came from his *seventh*-harmonic pattern of planets, and it was his passion to express his romantic inspiration through music which eventually enabled him to win through to the 'passionless calm' of the ninth harmonic. The harmonic charts must be looked at *in relation to each other* in order to understand the complexity of human behaviour.

CHAPTER 15
LINDA

In Chapters 15 and 16 we will look at two examples of how harmonics can be useful in astrological counselling. Unfortunately, I have not had time to do counselling since I started working with harmonic charts, and it is therefore necessary to take examples of clients who came to me before I started to use harmonic charts, and to discuss the ways in which the use of harmonic charts *would have been* helpful. We will call these two clients 'Linda' and 'Susan'.

Linda came to me for astrological advice when she was twenty-three years old. She had had an extremely wild youth, which she described in detail in a long written document which she sent to me. In childhood she had been very unhappy. Her father beat her mother, and was also a religious zealot. Linda cried a great deal, but was terrified of incurring her father's displeasure. She loved her mother but did not know how to relate to her: her mother was always asking for support from her children, which they felt unable to give. There were constant family quarrels.

But Linda was even more unhappy at school than at home. She felt that her teachers were always 'picking' on her. 'So I pretended every morning to have a stomach ache, in order to stay off from school. My mother gave in to this for a while but soon got wise to it and made me go, but I would return crying with some women who'd found me on the corner crying'. Eventually things improved, but Linda was never happy at school. 'I had many scenes with the teachers and could never get on with them. I also had very few friends: to gain friends I would do outrageous things in class, thus finally getting expelled from school.'

After leaving school, Linda left her parents and began 'crashing around' (to use her own phrase) in a world of hectic drink and sex. At eighteen she became pregnant. Her boyfriend left her, but she had the baby (a girl). She loved the baby, but was still

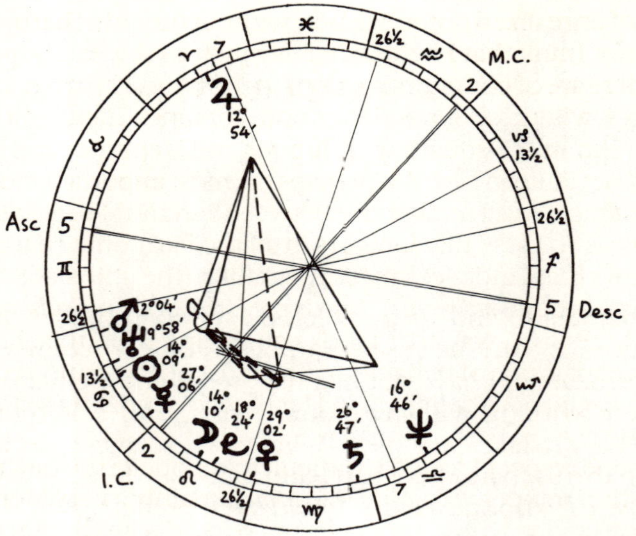

Linda: Radical chart

unhappy. 'Friends found me too heavy to have around, because I always had some problem I wanted to talk about and they said I brought them down'. After six months Linda had the baby fostered and went into a mental hospital, but discharged herself some months later on finding that there was no 'miracle cure'. She arranged to have the baby back as she felt that the baby was her future, but this did not work out: her sex life and her drinking had started again, and she found herself unable to give the baby as much love and care as it needed, so she had the baby fostered again. Later she moved into a 'group squat' and became involved with anarchist politics: 'I got in with an anarchist bomb-blowing group but left in the nick of time because I disagreed with their tactics – they are now serving seven years inside.' At the time that she consulted me she was very worried about 'night fears': 'I feel another presence in the room or I think that something is going to possess me'.

Such a case clearly presents an astrologer with problems. He must, at the very least, find factors in the birth chart which begin to explain Linda's behaviour and help her to make sense of herself. He must hope that he can go beyond this and guide Linda towards finding a way out of her problems by helping her to overcome her feelings of guilt and inadequacy and to find some kind of direction for her life. I definitely felt that Linda had a great deal of unrealized potential; but, without the aid of harmonics, I do not think that I helped her as much as she had hoped.

There are of course indications in the radical chart (p.174) of Linda's 'wildness': Sun and Mars both conjunct Uranus in a Water sign; Sun in a T-square with Jupiter and Neptune; and Moon conjunct Pluto in Leo. But perhaps the most important indicator is the almost exact semi-sextile between Sun and Moon: this semi-sextile is so close that it recurs in all the harmonic charts, and it shows that Linda seeks self-integration (the union of Sun and Moon) through *striving towards pleasure and enjoyment* (4 × 3). There is no way that Linda can escape from this: it is her dominant mode of behaviour. She must not be criticized for it but must learn to delight in it. Her life is essentially a search for pleasure.

In one respect I found it difficult to reconcile the birth chart with the facts of Linda's life. Linda was plainly a very Mercurial person (as one would expect from a person with Gemini rising, and Mercury in the third house close to the I.C.): she was highly articulate and communicative, and (in view of her limited education) her autobiography was extremely well-written: she had also written some very expressive poetry. Moreover it was clear that a large part of her life was 'lived in her mind' and that she was prone to thinking and theorizing about her own behaviour. However, it was also clear that her 'rationality' was of a very unorthodox and uncontrolled kind, and I found this difficult to reconcile with the fact that, in the radical chart, the only aspect to Mercury is a close sextile with Saturn. It was no good telling Linda that she was capable of great mental discipline, since this was plainly an attribute which she did not know how to find.

A full analysis of Linda's chart would of course be based mainly on the radical chart, drawing in insights from the harmonic charts where appropriate. However, since this book is concerned with harmonics, we will concentrate on the harmonic charts and will see where these yield additional insights which could not be gleaned from the radical chart.

H4 chart (p.176)

Linda's H4 chart contains little information that was not in the radical chart, but it emphasizes the pattern of Sun-Jupiter opposite Venus (visible as semi-squares in the radical chart) and

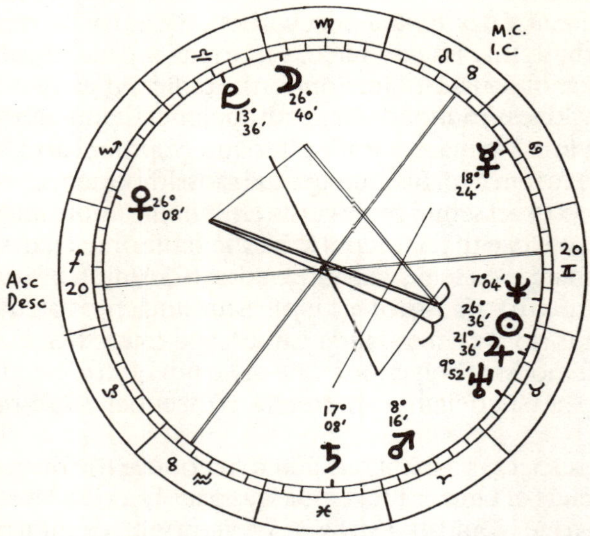

Linda: H4 chart

Sun trine Moon (which was a semi-sextile in the radical chart), and shows, through Moon sextile Venus, that this is all part of the one pattern of 'effort-behaviour'. Sun (in Cancer third house) and Jupiter (in Aries twelfth) are radically in square, showing that Linda's life is a struggle to overcome the obstacles to her growth and self-development; but the oppositions to Venus show that this struggle will tend to be *manifested* mainly through actions involving Venus. Among the interpretations which I have suggested for Venus-H4-Jupiter is 'striving to attract others by grand and impressive behaviour', and this is particularly appropriate for Linda as her Venus is in Leo on the fifth cusp. But beyond this she is searching for freedom through love, and love through freedom. She feels that she lacks these things but, as is common with oppositions in the H4 chart, she goes to such lengths to overcome this lack that she seems to have fully espoused the 'free love' which she feels she lacks.

The involvement of the Moon in this pattern of planets shows that, in her search for freedom and love, Linda is easily distracted by events in her environment. She tends to follow the easiest path towards pleasure, and so it is difficult for her to stick to a coherent plan.

And yet the Mars-Pluto opposition shows that Linda is very

conscious of this lack of 'stickability' and is doing her best to overcome it. She is willing to put up with great physical hardship and to push her physical resources to the limit in order to achieve her goals. But the lack of integration between Mars-Pluto and the other planets in the H4 chart has meant that Linda has not been able to use this 'stickability' to overcome her vacillations on the issues that really matter to her. Instead, it seems that the effects of the Mars-Pluto opposition have been: firstly, a lack of interest in physical comfort, which has enabled her frequently to change lodgings in pursuit of her immediate goals, without worrying about the squalor of her surroundings; and secondly, her attraction towards extremist political groups which have Mars-Pluto aims.

Thus the H4 chart is very indicative of the types of behaviour in which Linda has engaged. She has tended to see any new situation (such as, for instance, a new school) as a threat to her freedom and her capacity for growth, and the rest has followed from this. But this is not to say that Linda could not use this pattern more wisely. She has tended, for instance, to seek 'free love' through frequent changes of partner, because her insecurity has made it difficult for her to turn down any opportunity for an intimate relationship. On the basis of the H4 chart, I would be inclined to advise her to be more selective in her choice of partners (because Venus, if it is functioning properly, is always selective), and to seek a compatible partner who *in himself* can offer her 'freedom through love'. I would advise her also to use all her Mars-Pluto 'grit' to concentrate on this aim and to cope with the temporary frustration and loneliness which could result from such a policy. An understanding of the H4 chart could help Linda to discover what she really wants, and so help to overcome the obscure and all-pervading sense of guilt which she has felt about her own behaviour.

H5 chart (p.178)

Linda's H5 chart contains few new aspects (since the Sun-Moon quincunx, the Sun-Venus semi-square and the Mercury-Saturn sextile are carried over from the radical chart) and no very clear pattern, showing that Linda is not naturally a well-ordered or habit-bound person. By far the most important aspect in the H5 chart is the close Sun-Saturn conjunction, which explains much of the gloom and despondency which hangs over Linda's life in spite of her adventurous pleasure-seeking behaviour. Sun-

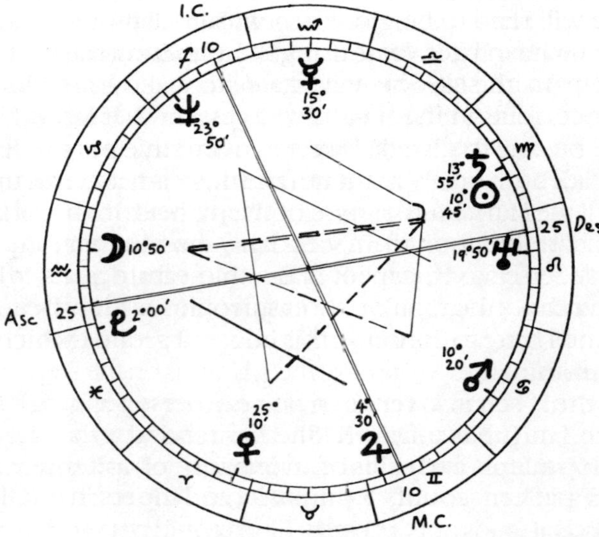

Linda: H5 chart

H5-Saturn means essentially that the person adopts habits of self-restraint, and it is clear that Linda, depsite her tendencies to emotionalism and exhibitionism, has always done this: in childhood, she told me, 'I never wetted the bed or sucked my thumb or had a favourite toy to comfort me: when I wanted to go for a piss I would always hold it in until the last minute'. But Sun-H5-Saturn can also mean that the person projects an image of himself as someone sunk in gloom and despondency, and this tendency is especially strong for Linda because Saturn is radically in the self-critical sign of Virgo and in the theatrical fifth house. Hence the 'heaviness' of which Linda's friends complained, and hence probably many of Linda's difficulties in making and keeping friends.

What can be done about this? My own view is that Sun-H5-Saturn does impart a certain inescapable seriousness, but that it does not necessarily lead to gloom and self-pity. By understanding this aspect, Linda can be helped to develop self-restraining habits which will help her to curb the wilder excesses of her own behaviour and so gain what she really wants. She has not done this until now, partly because she has not really known what she wants, but also because she has tended to see her self-restraint and seriousness as a *cause* of her problems rather than as a possible solution to them. But, like all Sun-Saturn aspects, this one will tend to be more beneficial in later life, and I can imagine Linda in later years using the Sun-Saturn aspect to get a firm grip on herself. She will be able to make good use of the experience of life which her earlier excesses will have taught her, and will be able to give her life a constructive purpose.

If Linda's birth time is accurate (and there is reason to suppose that it is), we can also take note of the planets on the Angles of the H5 chart: Pluto on the Ascendant, showing that she tends to portray herself to other people as ruthless and dedicated (Linda said in her 'autobiography': 'A lot of people see me as being very strong and I have to conserve this image: it seems that if I show a little weakness they won't like me'); Uranus on the Descendant, showing that she also tends to portray herself as original and eccentric; and Jupiter on the Midheaven, showing that she wants to develop talents and skills of an expansive Jupiterian nature. This H5 pattern could, I think, have helped in providing occupational guidance for Linda. She had been unemployed for long periods (in the days when unemployment was not so common as it is now), and she would have welcomed advice about her suitability for different types of job.

H7 chart (p.180)

It is when we turn to Linda's H7 chart that everything seems to fall into place. For Linda's H7 chart contains an extremely strong aspectual pattern involving nine of the ten planets, showing that she is by nature highly inspirational and romantic.

The pivotal planet in the pattern is Mercury, which is on the Ascendant and has no fewer than seven aspects. We can start with the Mercury-Jupiter conjunction which, if the birth time is right, is on the Ascendant and so will come across particularly strongly in her behaviour: also, its importance is increased by the radical positions of Mercury near the I.C. and of Jupiter in the twelfth house square to the Sun. In Chapter 10 I gave the following interpretation of Mercury-H7-Jupiter: 'Inspired by the idea of free and uninhibited thought and communication, leading to freedom and growth: free-ranging imagination: inspired to generosity and magnanimity: prone to exaggeration in communication (purple passages, ham-acting, overstating one's case)'. All this is clearly true of Linda. Perhaps she has tended to display the more negative sides of it (especially the exaggeration in communication), but I feel that if Linda can be

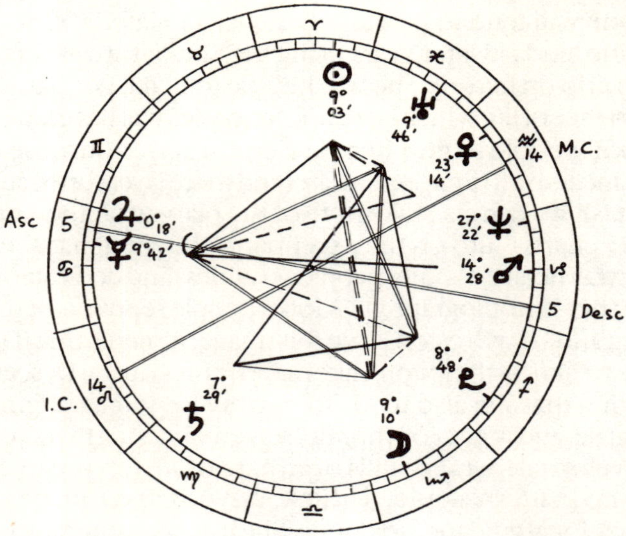

Linda: H7 chart

helped to overcome her guilt feelings and develop self-confidence, the warmth and generosity which are a part of Mercury-H7-Jupiter will (helped by the radical position of Mercury in Cancer) come more to the fore.

But also, Mercury in the H7 chart is involved in three other patterns:

square to the Sun, and semi-square to Venus
in a very close Grand Trine with Moon and Uranus
opposite Mars on the Descendant.

Firstly, Mercury square to the Sun in H7 shows that Linda is searching for an inspired theory about herself: and the semi-square to Venus shows that this search is connected with her search for love and for values and for her true femininity. The danger in this pattern is that Linda may tend to get carried away by unrealistic fantasies about the kind of person she is. She may, for instance, develop a picture of herself as a perfect mother, or as a slut, or as a long-suffering victim, or as a mental patient with suicidal tendencies, or as an anarchist's 'moll', and these self-theories will tend for a time to govern her actions and make it difficult for her to face reality. Yet Sun square to Mercury in the

H7 chart is potentially very creative: it gives the person a vision of himself which he may be able to communicate creatively to the world, and which may inspire him to action to turn the fantasy into reality. If Linda can exercise greater control over her self-theories (with the help of her radical Saturn sextile Mercury), she may find that they are a source of strength.

Secondly, the Grand Trine of Moon, Mercury and Uranus in H7 is a major source of the recklessness of Linda's behaviour. Moon-H7-Uranus can lead to great daring and foolhardiness, and Mercury-H7-Uranus leads to a desire for mental originality at all costs and to a desire to shock people by what one communicates. Linda is tremendously inspired by the idea of reckless behaviour and shocking communication, and she may tend to pursue it blindly, without thought for the consequences for herself or for others. Because these are trines and not squares, this is behaviour which comes naturally and effortlessly to Linda, and which brings her a great deal of pleasure. But it has clearly led her into situations which she has later found difficult and painful. One can only hope to make Linda more aware of this pattern, so that she has a clearer picture of what she is doing.

Thirdly, Mercury opposite Mars on the Ascendant-Descendant axis may well be an indication of Linda's 'night fears': 'I feel another presence in the room or I think that something is going to possess me: the feelings that one has before sleep seem sometimes to be of surrender to this "thing" that wants to possess me – so I end up fighting against sleep'. Mars on the H7 Descendant is projected out on to the world and becomes a fantasy of an external aggressor. This may well be symbolic of the emotional view which Linda has about men: she in her Mercurial rationality fighting against men in their raw Martian physicality and aggressiveness (note that radically Mars is in Cancer in the second house and so has qualities of enfolding possessiveness). Once again, one hopes that by pointing out to Linda the astrological causes of her fears one may be able to alleviate them.

Thus, we can see in the H7 chart the extremely wild, complex and emotional nature of Linda's ruling planet, Mercury. If we looked only at the radical chart, in which Mercury's only aspect is a sextile to Saturn, we would obtain a very misleading picture: the H7 chart is essential for an understanding of Linda's character.

This does not in fact exhaust the richness of Linda's H7 chart. There is also the close Sun-Pluto trine, the Moon-Saturn sextile, and the T-square of Saturn, Uranus and Pluto. The Sun-Pluto trine

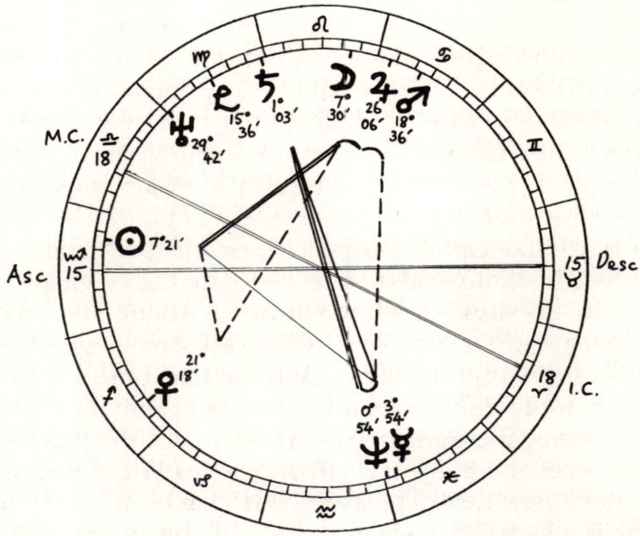

Linda: H9 chart

is a somewhat masochistic aspect: it implies that Linda is inspired by the idea of self-castigation and in fact obtains pleasure from it. At the time that I met her, Linda was involved in a very ruthless form of therapy, and it is clear that she would take easily to this and would find it a valuable outlet for her feelings. The Moon-Saturn sextile may increase the tendency to self-pity. And the Saturn-Uranus-Pluto T-square says something about the combination of moral qualities which inspires Linda: the Saturn-Uranus opposition is potentially a very creative aspect, and Pluto gives it extra force and determination.

When faced by a strong H7 chart, it is easy to suggest that because its content is fantasy, it should be suppressed: but clearly suppression would be neither possible nor desirable. Linda needs outlets for all her H7 energies, and in particular for the immense talent for creative and inspired communication which is contained in the H7 aspects to Mercury; and part of the aim of any consultation should be to explore with her the type of outlets which she can find. She certainly should continue to write poetry, and she should see it as part of her mission in life to communicate to the world her personal romantic vision.

H9 chart (p.182)

Only one aspect-pattern in Linda's H9 chart need detain us: this is the opposition between Mercury-Neptune and Saturn. The Mercury-Saturn opposition is in fact the sextile which was present in the radical chart, but the H9 chart shows that this sextile is closely linked with Mercury-H9-Neptune. The whole combination suggests: 'Mental discipline through rejoicing in the dissolution of one's individuality and in one's oneness with the human race'. Two years after I met her, Linda went to India and became a member of a religious sect, through which it seems likely that this objective was achieved. Maybe it should have been possible, from a study of the H9 chart, to predict that this was the path she would follow. The H9 chart often contains very strong clues on how a person can achieve an inner serenity which will help him to cope with life more successfully, and in Linda's case this message is particularly clear because of the strength of the Mercury-Saturn-Neptune aspect-pattern. By adding a spiritual dimension to her life, Linda will not become a different person, but we can hope that she will be better able to put her considerable talents to more creative, and less destructive, use.

Thus I believe that a study of the harmonic charts would have greatly helped me to understand Linda's personality and problems and to give her relevant advice.

CHAPTER 16
SUSAN

Susan came to me for astrological advice when she was thirty-five years of age. Her husband had recently left her, and she wanted advice on the directions which her life might now take. However, she was in a state of profound depression from which I was unable to rouse her. Whereas Linda had wanted to tell me everything about herself, Susan seemed to want to tell me nothing. She sat with an appearance of rock-like solidity, as though no force existed which could move her, and she replied monosyllabically to my questions and comments. She seemed to have no interests other than watching television, and no wish to acquire any other interests.

I regarded my attempt to help Susan as a failure, not only

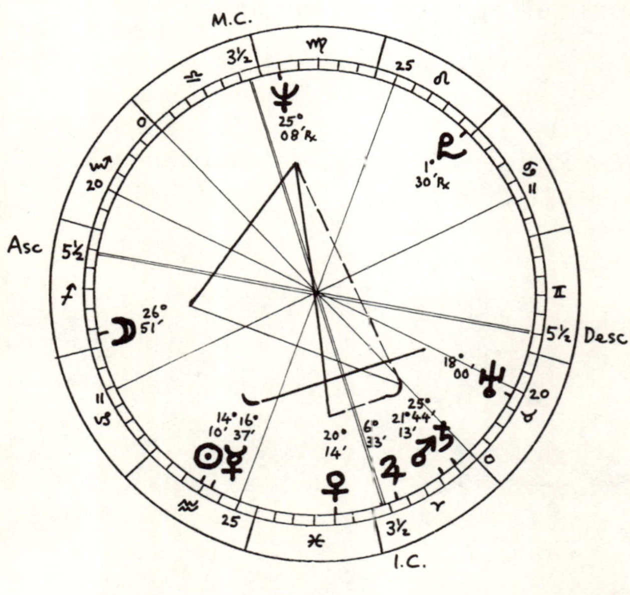

Susan: Radical chart

because of her lack of responsiveness, but also because I found it difficult to relate her chart (p. 184) to the person whom I saw before me. It was difficult to reconcile her rock-like immobility with her Moon and Ascendant in Sagittarius, or her lack of enterprise with her Mars, Jupiter and Saturn in Aries, or her apparently deep depression with her Sun and Mercury in Aquarius which should have enabled her to adopt a more detached attitude. Because of these features of her chart, I was led into making comments and suggestions which Susan found unhelpful and invalid.

As in Linda's case, we will look at Susan's harmonic charts to see whether they provide additional insights which would have solved the 'mystery' of Susan's behaviour, and would have enabled a more valid and helpful interpretation to be made. In this case, however, our conclusions will have to be speculative, as I know so little about Susan.

H4 chart (p. 186)

The H4 chart contains one very important aspect, which is the Sun-Mars square. This aspect is probably the biggest single clue to Susan's behaviour.

In general it can be said that people with Sun-Mars squares in H4 are conscious of a lack of the force principle within themselves, and are striving hard to rectify this lack. They may develop a life-style in which the force principle is very strongly manifested, although they will tend to do this in a tense and nervous way, as though trying to prove something about themselves. But it is probably easier for men than for women to work out this aspect in a positive way. Women with this aspect may well be unable to accept Mars as part of their own personality, and will project Mars on to the men in their lives: and this is especially true for Susan, since Mars is in Aries in conjunction with Saturn and so represents a strong and dominant masculinity, which she cannot accept in herself without renouncing her own very soft femininity (Venus in Pisces). Hence she will tend (as so often with squares in the H4 chart) to regard Mars as the 'enemy' against whom she must fight in order to find her own true Sun-nature. But since Mars belongs to the enemy rather than to herself, she cannot fight him by Martian methods, so she can fight him only by passive resistance. (Plainly she was adopting this tactic with me: she saw me as trying to force her to change her behaviour, and so she resisted me passively. A woman astrologer might well be more successful with Susan than I was.)

The other aspects to Mars in the H4 chart reinforce this point. Since Mars is opposite to the Ascendant-Descendant point, she will tend to see it as the force which prevents her from relating to the world; and, since Mars is sextile to Jupiter, she will tend to see the capacity for freedom and growth as something which belongs to the 'enemy' rather than to herself. Hence her tendency to opt out of relating to the world, and to spend her life in front of the television set. And finally, the Venus-Mars trine (which is a semi-sextile in the radical chart) is an indication of how deeply Susan *needs* the Mars principle (even though she cannot accept it as part of herself) and of how deeply wounded she must have been by her husband's departure: for she is seeking pleasure through the reconciliation of Venus with Mars, and she can find Mars only in a man.

But I think it is possible that Susan can learn to use this aspect more positively. If an astrologer or counsellor can break down her resistance, and also if she spends a long period without a man and so is forced to become more self-reliant, she may well be persuaded to take up an activity within which she can manifest the force principle in the tense and dramatic manner which is

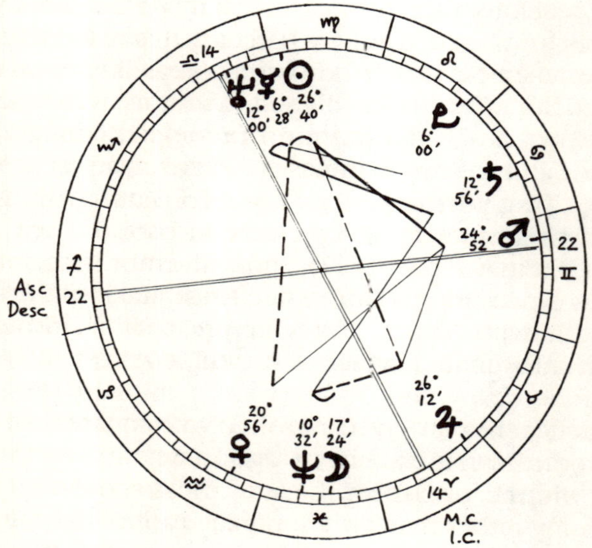

Susan: H4 chart

typical of Sun-square-Mars in H4. Judo is the particular activity which occurs to me (although there are probably others which would do just as well), and I think that such activities would be immensely beneficial for Susan, since they would help her to accept Mars as part of herself. Even keep-fit classes would be helpful. (H4 problems are *action*-problems, and can only be solved by means of actions and activities.)

Also in the H4 chart there is the Saturn-Uranus square, which represents a dilemma between discipline and independence. Since Uranus is conjunct Mercury it represents mainly *mental* independence, and since Saturn is radically conjunct the 'enemy' Mars it represents the restrictions imposed by her husband, so that she probably accepted physical restrictions while refusing to be influenced by him mentally. Following her usual pattern of passive resistance, this refusal probably took the form of constantly listening to the television rather than to him. Now that her husband has left, she is free to seek a more genuine mental independence and originality, and again one could advise her on activities which might help to bring this about.

H5 chart (p.188)

In Susan's H5 chart we are struck immediately by the close Sun-Saturn conjunction, and by the fact that this is the same aspect as that which dominated Linda's H5 chart. It may seem strange to find the same aspect in the charts of people as different as Linda and Susan. But, whereas Linda's Sun-H5-Saturn was an almost isolated aspect, Susan's is the centre of a complex configuration of planets: Sun and Saturn opposite to Venus and trine to Moon and Pluto. Thus, whereas Linda's Sun-H5-Saturn behaviour, though strong, seemed to be a separate and isolated part of her character, Susan's is part of a far more all-embracing pattern. Susan, like Linda, adopts habits of self-restraint, but they are a far more dominant feature of her character than of Linda's, and have helped to turn her into a person who does very little with her life.

The exact Sun-Venus opposition in H5 means that Susan tries to attract other people in stylized ways. There is a problem about sincerity with such an aspect: the person is so concerned about the *style* of his behaviour (am I doing this in the right way? will they approve of my behaviour?) that it becomes difficult to behave naturally and to express feelings sincerely. In Susan's case the involvement of Saturn makes things still more difficult:

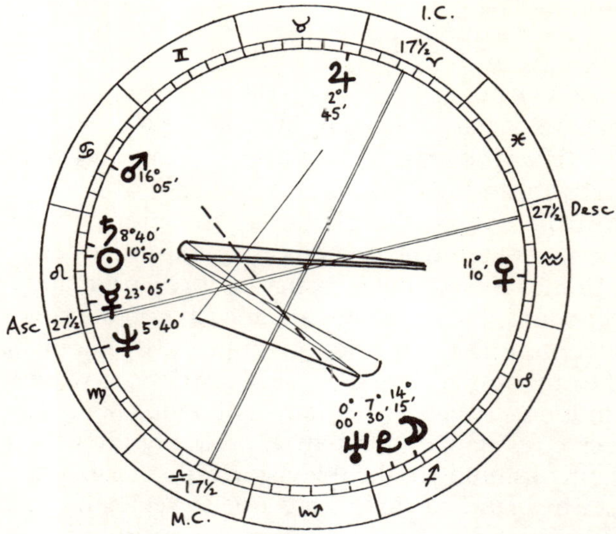

Susan: H5 chart

Venus-H5-Saturn is concerned with trying to attract others by one's restraint and devotion to duty, and with a belief in the value of discipline and control. Thus Susan becomes a person who finds it very difficult to show other people how she feels about them. The natural warmth and openness of Venus in Pisces is crushed under the weight of the detached Aquarian Sun and the harsh Saturn in Aries, and probably becomes diverted into its negative function of repelling people whom she does not like, rather than attracting those that she does.

It is not going to be easy to change Susan in this respect, since the Sun-Moon trine shows that Susan naturally enjoys behaving in ordered and stylized ways. She is naturally a person who likes to be dominated by habits and routines, and she will not be at ease if she tries to adopt a more experimental approach to life. And the involvement of Saturn and Pluto in this trine imposes on it its own particular flavour of discipline and dedication. In Chapter 10 I said about Moon-H5-Pluto: 'In many cases the person seems to be bent on self-destruction (physical or mental) through his habits of behaviour: in other cases there is a facade of hardness and ruthlessness'. Susan's Pluto is radically unaspected in the eighth house, so it is likely that Pluto (like Mars) is projected on to the outside world and is interpreted by Susan as the ruthlessness of other people in dealing with her. Yet Pluto in the radical chart is trine to the Ascendant and sextile to the M.C., so it is unlikely to have brought her much distress. Thus the Sun-Moon-Saturn-Pluto trine in H5 creates a picture of Susan as a happy masochist, a person who enjoys being enslaved, and who carries out her duties as a slave punctiliously to the point of damaging her own health. On suddenly receiving her unwanted freedom, it is natural that such a person should feel that she is in a foreign country where she cannot speak the language and has no clue how she should behave. She will be likely to cling tenaciously to the habits which have served her in the past, rather than experimenting with styles of behaviour which are more appropriate to the new situation.

Yet Susan's thinking mind (Mercury) stands apart from this configuration. With both Mercury and Neptune near the H5 Ascendant (if the birth time is right), Susan will want to portray herself to other people as thoughtful and communicative and also as sympathetic and caring. But the patterns of behaviour imposed by the Sun-Saturn-Venus-Moon-Pluto configuration may leave her little room to do so. In so far as she is locked into H5 patterns of behaviour, her detached Aquarian mind is able to stand aside from this behaviour, observe it coolly, and think it out afresh. This seems to offer some hope that Susan will be able to adapt successfully to the new situation.

H7 chart (p.190)

Susan's H7 chart presents a very different picture. Firstly, Sun-H7-Jupiter suggests that Susan is inspired by the idea of her own growth and freedom. Since Jupiter in the radical chart is close to the I.C., I feel that this is probably connected with a desire to recover the sense of freedom which she feels she had in the past, probably in her childhood. But, since Jupiter is in the fourth house, it may also be expressed as a desire to build a home within which she and her family can feel free and untrammelled. But Sun is also square to Pluto, which implies that she is also seeking inspiration in dedication and self-punishment. With Sun linked to Pluto both in H5 and in H7, the Plutonian urges will be very strong. She will settle into Plutonian routines, but there will also be times of H7 inspiration when she will drive herself even harder, with a feeling of emotional commitment.

There are two other strong configurations in Susan's H7 chart: Moon square Uranus and semi-square Venus, and Mercury

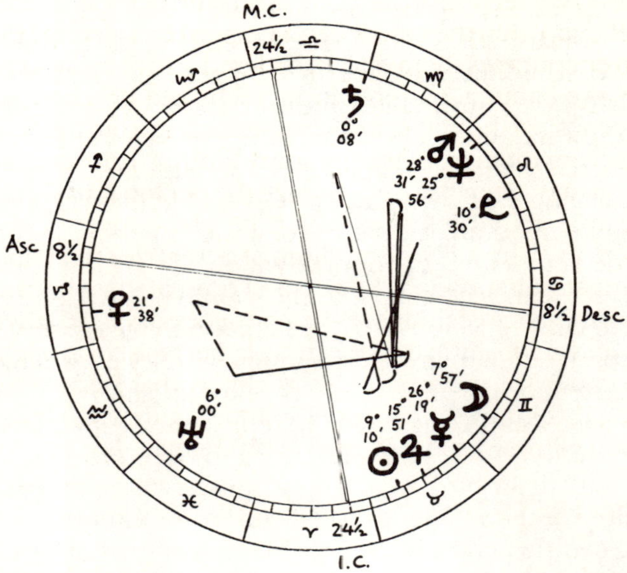

Susan: H7 chart

square Mars and Neptune. The Moon-Uranus square shows that Susan *wants* to find inspiration and excitement in behaviour that is the opposite of her usual pattern – vivid, striking, electrifying, daring and foolhardy behaviour: and the semi-squares to Venus seem to show that she feels it is her feminine qualities (perhaps the very fact that she is a woman) which prevent her from displaying this type of behaviour. The Mercury-Mars-Neptune pattern (the importance of which is increased by the radical position of Neptune near the M.C., showing that Susan is striving towards Neptunian goals) shows that she is inspired by the idea of idealistic programmes of action, and wants to use her rational mind to develop these programmes and to communicate to the world about them.

Thus the H7 chart presents a complex picture which seems to be out of tune with the behaviour which Susan exhibited when she met me, and it would be necessary to question her to find out the ways in which she has manifested the tendencies of this chart. But it seems at least possible that the H7 pattern (or at least the Moon-Uranus and Mars-Neptune-Mercury components of it) has not manifested strongly in her behaviour, but has remained at the level of unfulfilled dreams and fantasies, of which Susan may have been only partly conscious, but which may have increased her feeling of dissatisfaction with her way of life. This is, I believe, often the case when a person has a strong H5 chart which encourages them into an ordered and habit-bound way of life: the habitual patterns induced by the H5 chart act as a defence against the wilder impulses stemming from the H7 chart, and prevent them from reaching any kind of fruition. The best solution may often be to encourage the person into some kind of creative activity within which he can convey a creative *message* (H7) through the medium of a creative *style* (H5), thus fulfilling simultaneously the needs of both of these harmonics. But in Susan's case I think it would be necessary to lead her very gently and gradually in this direction, as her resistance might well be strong. In the meantime it seems likely that Susan is finding a kind of fulfilment for her H7 fantasies through vicarious emotional involvement in television dramas.

H9 chart (p.192)

The most noticeable feature of Susan's H9 chart is the T-square involving Moon, Jupiter and Mercury. When we look back at the radical chart (p.184), we see that Mercury is almost exactly at the midpoint of Moon and Jupiter. Ebertin[1] interprets Mercury at the Moon-Jupiter midpoint as meaning, 'Far-reaching plans, a large sphere of thinking, consciousness of objective, the urge to learn and to study'. These are not qualities which we would naturally associate with Susan. But the H9 chart shows that this midpoint has the quality of H9, or more strictly of H36 since this is the harmonic in which these three planets come together into a conjunction: and H36 is to do with *striving towards joy*. So we can say that Susan's route towards happiness and joy lies in striving towards the widening of her interests and the opening-up of her consciousness. She needs to try to free her mind from restrictions and to make herself more receptive to the influence of the environment. Moon-H9-Jupiter means a delight in the sheer marvellousness of the world and the richness of the opportunities which it offers: Susan needs to seek this kind of joy, and Mercury shows that she should also try to convert it into words and to communicate her joy to people around her. The radical Moon trine Saturn (which reappears as a conjunction in the H9 chart) shows that she would do this cautiously, without losing sight of reality. And the radical positions of Moon, Mercury and Jupiter in the first, second and fourth houses suggest that she would be trying to make sense of herself-in-her-private-

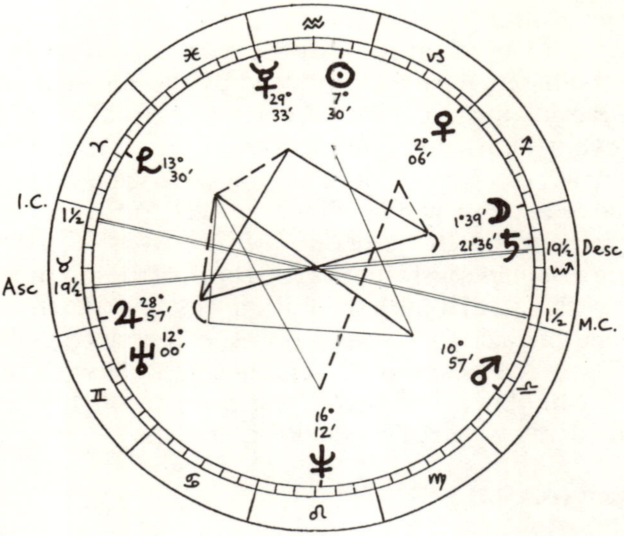

Susan: H9 chart

world, rather than trying to investigate wider mysteries (though Moon radically square Neptune in the ninth house works against this).

Perhaps it would be difficult to convey all this to Susan in a practical, helpful way. But the H9 chart does, I believe, tell us something about how we would feel if we could forget about all our conscious worries and strivings and concentrate on our sense of 'being-ness' in the world at the present time. Thus, the H9 chart can, in a sense, offer solutions to problems. But an astrologer who concentrates too much on the potentialities of the H9 chart may well be accused by the client of ignoring the real problems and of being impossibly unrealistic and optimistic. The astrologer must be able to judge when the client is ready to hear about his H9 chart.

Susan's H9 chart also shows Mars trine to both Sun and Uranus, and opposite to Pluto. This seems to show that Susan is capable of intense joy in physical activity and in the consciousness of her own forcefulness and dynamism. But we must not forget that Mars is the planet which (as the H4 chart showed) Susan tends to project on to other people (and especially her husband) rather than accept as part of herself. As long as Mars is so projected, this H9 configuration can only mean 'joy in union with a strong man', and so detracts from Susan's self-sufficiency and from her ability to cope with separation and divorce. Thus we come back to the need to find ways of persuading Susan to accept Mars as part of herself. This is, I believe, the most important clue to the type of help and therapy which Susan needs. Her harmonic charts show that once this has been achieved, there are many ways in which she can develop her life and achieve greater happiness and self-fulfilment.

I am of course aware of the limitations of these analyses of Linda's and Susan's harmonic charts. These limitations are due to a number of factors: firstly, my personal limitations; secondly (especially in Susan's case), the limits of my knowledge of the clients' actual life-patterns and of their feelings and aspirations; and thirdly, the limits of our understanding of the working of harmonic charts. Nevertheless, I believe that the cases show that the harmonic charts can provide the astrologer with a number of vital clues which can lead to fuller understanding and more effective counselling.

PART FOUR: OTHER APPLICATIONS

In Part Four, we shall look at some other ways in which astrologers may be able to use harmonics in their work. Most of these are completely new approaches, and have not been fully developed and tested either by myself or (so far as I know) by any other astrologer. All we can do, therefore, is to make some suggestions about lines of exploration that seem to be interesting and promising.

CHAPTER 17

HARMONIC SYNASTRY

First, we shall look at the possibilities of using harmonics in synastry. When astrologers are studying a relationship between two people, they customarily note the aspect-links between planets in one chart and planets in the other chart. Therefore, if we have drawn up a series of harmonic charts for each of the partners in the relationship, it is natural that we should also look at the aspect-links between planets in each of the pairs of harmonic charts (i.e. between 'his' H4 chart and 'her' H4 chart, 'his' H5 chart and 'her' H5 chart, and so on). This means simply that we are looking in greater depth at the pattern of aspectual links between the two birth charts.

The method which I have used is as follows:

1 Draw up the radical, H4, H5, H7 and H9 charts for each of the partners in the usual way. (As in any other type of synastry, it is desirable to study and analyse the charts of each of the partners separately, before going on to analyse the nature of the relationship between them.)
2 List the aspect-links between planets in each of the pairs of charts, using harmonic orbs. The orbs which I use in synastry are *half* of those used in ordinary natal work (see Appendix III). This means that the orbs used in synastry for aspects between planets are as follows:

	Ordinary aspects	*'Close' aspects*
Conjunction	6°	1°
Opposition	3°	30′
Trine	2°	20′
Square	1°30′	15′
Sextile	1°	10′
Semi-square	45′	7′
Semi-sextile / Quincunx	30′	5′

3 List also the aspect-links between planets in one chart and the Angles (Ascendant, M.C.) of the other chart, and between the Angles of one chart and the Angles of the other chart. Unless the birth times are very exactly known only conjunctions and oppositions should be noted in the H5, H7 and H9 charts; but in the H4 chart squares and trines may also be noted. The suggested orbs for aspects to the Angles are as given below. (The higher figures for oppositions and sextiles are based on the fact that, except in the H4 chart, oppositions to the Ascendant/M.C. are conjunctions to the Descendant/I.C., and sextiles to the Ascendant/M.C. are trines to the Descendant/I.C.)

	Radical chart	*H4 chart*	*H5, H7 & H9 charts*
Conjunction	6°	6°	6°
Opposition	6°	3°	6°
Trine	2°	2°	
Square	1°30′	1°30′	
Sextile	2°		
Semi-square	45′		
Semi-sextile } Quincunx }	30′		

4 Enter the aspects on to a *matrix* (or grid), with the planets and Angles of one partner on one axis of the matrix, and the planets and Angles of the other partner on the other axis. A separate matrix should be drawn for each pair of harmonic charts.

We can now interpret the meaning of the links between the partners, as shown by the aspects in the harmonic matrices.

Morecambe and Wise

Three examples of this technique will suffice. The first is the relationship between Eric Morecambe and Ernie Wise, two comedians who depend on each other for their success, who have developed a distinctive style of acting together, and who are able to inspire each other to wit and lunacy. They do not write their own material, but they impose their own style on it, no matter who the scriptwriter may be. Morecambe usually appears as the aggressor and Wise as the innocent victim (though in their earlier work these roles were often reversed). Unlike many

ERIC MORECAMBE

ERNIE WISE

H1	☉	☽	☿	♀	♂	♃	♄	♅	♆	♇	A	M
☉												
☽			☌			*						
☿		☍				*̲						
♀	△								⚻		⚻	
♂		∠			△							☍
♃	△										⚻	
♄							☌					
♅							△		⚻		⚻	∠
♆						☍			☌		☌	
♇										☌		
A				☍						□		
M										☌		

H4	☉	☽	☿	♀	♂	♃	♄	♅	♆	♇	A	M
☉												
☽						△						
☿						△̲						
♀						∠			△		△	
♂		☍	⚺									
♃											△	
♄												
♅									△		△	☍
♆						☌		∠				
♇									⚻	☌		
A										☌		
M												

H5	☉	☽	☿	♀	♂	♃	♄	♅	♆	♇	A	M
☉						∠		⚻				
☽			⚺		△̲	*	∠					
☿			□		☌̲	*̲						
♀												
♂												
♃												
♄										□		
♅												
♆												☍
♇			☌	△̲								
A					☌							
M								☍				☌

H7	☉	☽	☿	♀	♂	♃	♄	♅	♆	♇	A	M
☉		∠							☌	□	☌	
☽				△						☍		
☿				☌̲		*						
♀				☍						△̲		
♂									☍		☍	
♃				☍̲		△						
♄						∠̲						
♅				△		☍				*̲		
♆					□							
♇	☌̲						☍					
A									☌		☌	
M												

H9	☉	☽	☿	♀	♂	♃	♄	♅	♆	♇	A	M
☉		☌			△							
☽						☍						
☿						☍̲						
♀			☍̲					△				
♂						△						
♃			☍									
♄												
♅			☌̲									
♆												
♇					*̲							
A	☍						☌					
M												

Fig. 1 Morecambe and Wise: Harmonic matrices

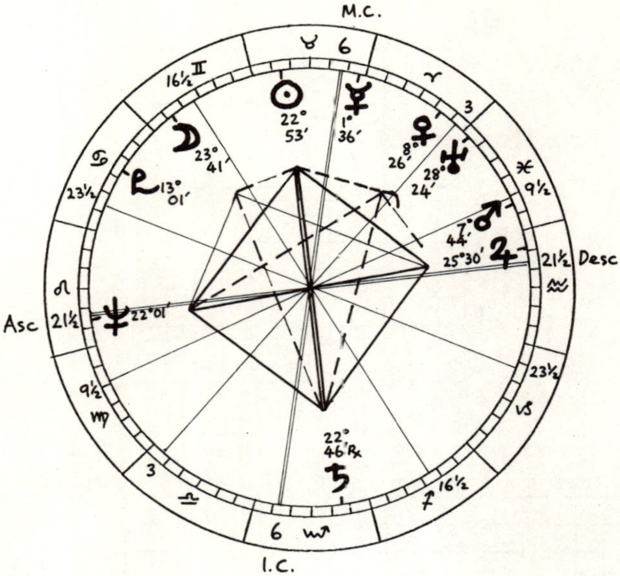

Eric Morecambe: Radical chart

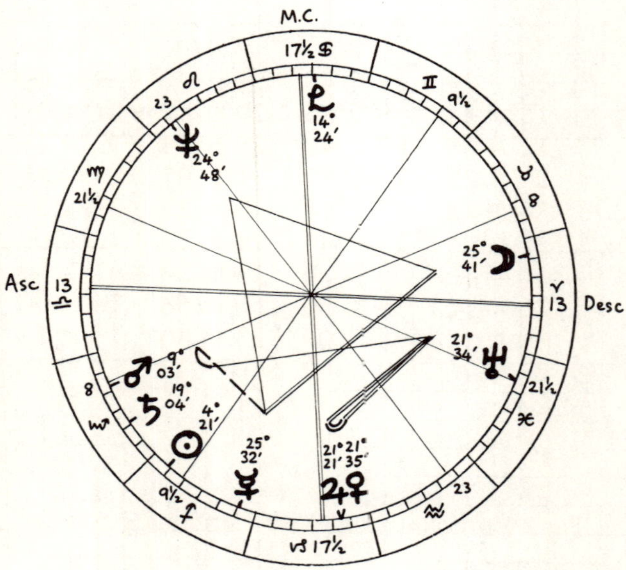

Ernie Wise: Radical chart

Eric Morecambe

	H4	*H5*	*H7*	*H9*
☉	1 ♏ 32	24 ♐ 25	10 ♈ 11	25 ♋ 57
☽	4 ♓ 44	28 ♉ 25	15 ♏ 47	3 ♉ 09
☿	6 ♌ 24	8 ♍ 00	11 ♏ 12	14 ♑ 24
♀	3 ♉ 44	12 ♉ 10	29 ♉ 02	15 ♊ 54
♂	0 ♑ 56	8 ♐ 40	24 ♎ 08	9 ♍ 36
♃	12 ♏ 00	7 ♎ 30	28 ♋ 30	19 ♉ 30
♄	1 ♏ 04	23 ♊ 50	9 ♎ 22	24 ♑ 54
♅	23 ♓ 26	22 ♓ 00	18 ♓ 48	15 ♓ 36
♆	28 ♎ 04	20 ♓ 05	4 ♑ 07	18 ♎ 09
♇	22 ♉ 04	5 ♍ 05	1 ♈ 07	27 ♎ 09
A	26 ♎	17½ ♓	0½ ♑	13½ ♎
M	24 ♌	0 ♎	12 ♐	24 ♒

Ernie Wise

	H4	*H5*	*H7*	*H9*
☉	17 ♐ 24	21 ♌ 45	0 ♑ 27	9 ♉ 09
☽	12 ♋ 44	8 ♌ 25	29 ♍ 47	21 ♏ 09
☿	12 ♓ 08	7 ♐ 40	28 ♉ 44	19 ♏ 48
♀	26 ♊ 20	17 ♈ 55	1 ♐ 05	14 ♋ 15
♂	6 ♍ 12	15 ♈ 15	3 ♋ 21	21 ♍ 27
♃	25 ♊ 24	16 ♈ 45	29 ♏ 27	12 ♋ 09
♄	16 ♎ 16	5 ♊ 20	13 ♍ 28	21 ♐ 36
♅	26 ♒ 16	17 ♒ 50	0 ♒ 58	14 ♑ 06
♆	9 ♏ 12	4 ♈ 00	23 ♑ 36	13 ♏ 12
♇	27 ♉ 36	12 ♍ 00	10 ♈ 48	9 ♏ 36
A	22 ♉	5 ♐	1 ♑	27 ♑
M	10 ♊	27½ ♍	2½ ♉	7½ ♐

Fig. 2 Morecambe and Wise: Harmonic positions

'double acts' they work on a basis of equality, and there is a close bond of affection between them, although they lead separate lives outside their work. They are also remarkable for being able to attract other people (including 'serious' actors such as Glenda Jackson, and even politicians such as Harold Wilson) into their act. The attraction seems to be that appearing on the Morecambe and Wise show and agreeing to look ridiculous is a way of showing one is warm-hearted and un-pompous.

The matrices of aspects between each of the pairs of charts (i.e. between Morecambe's radical chart and Wise's radical chart, Morecambe's H4 chart and Wise's H4 chart, and so on) are shown in Figure 1 (p.199). Also shown are Morecambe's and Wise's radical charts (p.200). For reasons of space I have not included the harmonic charts, but the positions of the planets and Angles in each of the harmonic charts are shown in Figure 2 (p.201), and readers will be able to draw the charts if they wish to follow this case in detail.

Radical chart

Perhaps the most important link between Morecambe's and Wise's radical charts is the close link between Morecambe's Jupiter and Wise's Moon, Mercury and Neptune. M's Jupiter is sextile to W's Moon and Mercury (being at the midpoint between them), and these links are so close that they are carried over into the harmonic charts; also, it is opposite to W's Neptune. Thus, M's Jupiter is linked in with the Grand Trine of Moon, Mercury and Neptune which is the most striking feature of W's radical chart and which signifies his pleasure in communicating his feelings and achieving empathy with others (had Wise not become a comedian, he might have been an excellent counsellor or therapist). The importance of this link is increased by the fact that M's Jupiter is in his seventh house and so represents what he is seeking in relationships; and it is also one of the points of the Grand Cross of Sun, Saturn, Neptune and Jupiter which dominates M's radical chart. M is, as the Grand Cross shows, very tense and highly-strung (in contrast to W's far more easygoing nature), but he seeks relief from the tension through expansive relationships. But W's Moon is also in *his* seventh house, and so represents what *he* is seeking in relationships. Thus M and W are drawn to each other especially through the sextile between M's Jupiter and W's Moon. The sextile, which represents 'searching for pleasure', is especially appropriate for a relationship whose purpose is to create laughter.

Also we see that M's Moon and Mercury are linked with W's Moon and Mercury, so that the relationship is very much to do with the articulation of perceptions, feelings and emotions.

The trine between M's Sun and W's Venus and Jupiter (which are in close conjunction with the I.C.) shows that M feels naturally in tune with W's open-mindedness and warm-hearted generosity; and the Mars-Mars trine shows a relationship within which uninhibited energy can be cheerfully expressed.

H4 matrix

The H4 matrix shows few new aspects, indicating that the relationship between M and W is not primarily one of conflict and effort. The most important link is the set of trines (which were shown as quincunxes in the radical matrix) between M's Neptune-Ascendant and W's Venus-Jupiter-Uranus. This shows that M's idealistic and 'ungrounded' way of relating to the world (Neptune rising means, perhaps, that the person comes across as someone who does not really care about himself) is linked in with W's warm-hearted and yet self-centred generosity. But the link is a tense one (having the nature of 'striving towards pleasure'), and the effects of this are evident in the unresolved tension in the Morecambe and Wise shows between W's air of self-satisfied friendliness and M's air of destructive fantasy.

The opposition between M's Moon and W's Mars is important, since H4 oppositions tend to result in a strong pattern of manifestation. In Chapter 10 we interpreted Moon-H4-Mars as meaning 'effort to overcome the inertia which seems to surround one, and to arouse others to action'; and M does often seem to be trying to rouse W from inertia. (H4 links are always concerned with an effort to overcome a perceived lack: thus, from M's point of view, the effects of this aspect would be seen as 'He seems to be inactive, I must try to make him act', whereas from W's point of view they would be seen as 'He seems to be unresponsive to my actions, I must try to make him respond'.)

The close semi-square between M's Venus and W's Sun represents another potential area of conflict (perhaps in the form of W feeling belittled and 'put down' by M's attempts to draw attention to himself and make himself seem lovable). Since this is a semi-square rather than an opposition, we would expect it to result less in open action and more in unexpressed resentment. Nevertheless, M and W seem to have found a way of expressing it in their act, in which W often seems humiliated by M's aggressive (Venus in Aries) attention-seeking behaviour. The semi-square between M's Jupiter and W's Venus represents a reverse pattern, in which M feels that his freedom is curtailed by W's attempts to steal the limelight.

H5 matrix

The H5 matrix shows an extremely strong pattern of aspects linking M's Mercury and Mars with W's Moon and Mercury (with M's Jupiter also involved through the sextiles which are carried over from the radical matrix). Of the six aspects linking these planets, three are underlined as being 'close'. Thus, this is a very powerful pattern which, because it is in H5, is concerned with the *style* in which M and W relate to each other. With these close links between their 'personal' planets, M and W are easily able to create a *style of interaction* which can become the hallmark of their relationship.

The close square between the two Mercuries shows that their style is essentially to do with communication: it is a verbal style. But the square shows that (within their deliberately created style) they communicate tensely, seeming always to be at cross purposes with each other. (The way in which M communicates is affected by his Mercury conjunct Pluto and square Mars in his H5 chart; the way in which W communicates is affected by his Mercury opposite Saturn and trine Moon and Neptune). And the conjunction between M's Mars and W's Mercury shows that (still within their stylized behaviour) M tends to dominate W, while W tries to reason with M. But the cheerfulness and humour of the relationship is shown especially by the close trine between M's Mars and W's Moon. In Chapter 10 we defined Moon-H5-Mars as meaning, 'Adopting a style of quick, dynamic and forceful response to events, and of open, direct and powerful communication: may be connected with knock-about humour'. This is the nature of M and W's stage relationship (with the added pleasure and enjoyment brought by the trine); but it is mainly M who provides the force and dynamism, and W who responds to it.

It is interesting to look at this H5 matrix alongside M and W's separate H5 charts. The most striking feature of M's H5 chart is Sun/Saturn square Uranus (with Sun opposite Saturn carried over from the radical chart), showing that M is a person who strives very hard to develop a highly precise and original style of behaviour to cover up his own insecurity. But this pattern finds little response in W's chart: it is as though M does not need W's help to develop this aspect of his style, he can do it well enough on his own. The configuration of Mercury, Mars and Jupiter in M's H5 chart is apparently less important, but it is this configuration which meets the strongest response from W. Similarly, in W's H5 chart, the most striking feature is the conjunction of Venus-Jupiter with Mars, but this finds little response from M; it is the Moon-Mercury trine in W's H5 chart to which M responds most strongly. Thus we can see how each of them helps to develop the other's talents in certain ways, but not in others: with a different partner, each of them could have emphasized a different side of his talents and developed a different style.

Also in the H5 matrix, we can note the Sun-Sun trine, which shows that at the deepest level M and W gain pleasure and enjoyment from interacting with each other in stylized ways: and (if the birth times are accurate) the Midheaven-Midheaven conjunction, which seems to show that their paths come together through the development of a joint style.

H7 matrix
The H7 matrix shows an exceptionally strong pattern of links between planets. The planets mainly involved are M's Venus, Jupiter and Pluto (which are linked together by trines and sextiles in M's H7 chart) and W's Moon, Mercury, Venus, Jupiter and Uranus (which are linked in a very strong kite-formation in W's H7 chart).

This pattern is extremely complex and I will not attempt to analyse it in detail. In general, however, it shows that M and W inspire each other: they help each other to get into an H7 mode of behaviour which, of its very nature, is impulsive, unpredictable and intuitive. The H5 links set the *style* of their interaction; the H7 links determine its inspirational *content*. But the complexity of the H7 links suggest that M and W are able to inspire each other in very varied ways: hence the diversity and richness of their joint performances.

Since Venus and Jupiter are involved for both partners (with a Venus-Venus opposition and a Jupiter-Jupiter trine), we would expect their mutual inspiration to be partly of a Venus-Jupiter nature. In Chapter 10 we described the meaning of Venus-H7-Jupiter as follows: 'Romantically moved by ideas of freedom, growth and progress, believing in uninhibited and unconditional love: inspired to grand and extravagant displays of feeling: seeing beauty in whatever is grand, expansive and uplifting: a romantic optimist'. This might seem to imply that M and W are romantically attached to each other: and indeed I think this would be the case if either M or W was a woman, or if they both had homosexual tendencies. As it is, I believe that the bonds of affection between M and W are very strong, and that each of them feels liberated by his relationship with the other. They are certainly able to express their feelings to each other in 'grand and extravagant' ways. Moreover, the theme of 'romantic optimism' corresponds closely to the 'message' which emerges from the Morecambe and Wise shows: their theme-song is 'Bring me sunshine', and (despite their surface rudeness) they spread around them an atmosphere of warmth and kindness which helps to account for their popularity, and also for the eagerness of other actors to appear in their shows.

Thus, both the H5 matrix and the H7 matrix are essential for a full understanding of the Morecambe-Wise partnership. Without the close H5 links, the warmth and inspiration shown in the H6 matrix could not be welded into successful theatrical routines: but without the H7 links, the partnership would be too much a matter of form, and the tension inherent in their style (Mercury square Mercury in H5) might well cause a break-up of the partnership.

H9 matrix

The links shown in the H9 matrix indicate the kind of happiness and peace which M and W are able to find in each other's company. Links of this type could be very important in a love-relationship, but their effects are less easy to discern in a relationship (such as that of M and W) whose success depends on activity, style and inspiration.

Nevertheless, there are some interesting links in the H9 matrix. Firstly, M's Mercury is closely linked with W's Venus, Jupiter and Uranus: these three planets were linked (Venus-Jupiter sextile Uranus) in W's radical chart and signify W's natural warmth and independence of spirit. Thus, the H9 link between these planets and M's Mercury shows that M feels that he understands this side of W's character and obtains intellectual satisfaction from it: and also W gains pleasure from the feeling that M understands him at this level.

Secondly, M's Jupiter is linked with W's Moon, Mercury and Mars. The oppositions to Moon and Mercury are carried over from the radical matrix (where they appeared as sextiles), but the Jupiter-Mars trine is new in the H9 matrix. In order to interpret this, we must remember that (as we said before) M's Jupiter is radically in his seventh house and shows that he is seeking relief from his tensions through expansive relationships. The H9 trine with W's Mars helps him to feel that in W's company he can relax and 'spread his wings'.

Thus, although the H9 matrix may add little to our understanding of the Morecambe-and-Wise comic partnership, it provides some additional insights into the relationship between the two men.

Scott and Zelda Fitzgerald

Having dealt with Morecambe and Wise in some detail, we will now discuss two other cases more briefly, concentrating only on a few outstanding points from the harmonic matrices. First, we will take the case of the writer F. Scott Fitzgerald and his wife Zelda. Their harmonic matrices are shown in Figure 3 (p.209), and the positions of the planets in their radical and harmonic charts are shown in Figure 4 (p.210).

Scott and Zelda Fitzgerald together epitomized the Jazz Age. According to Rodden they 'were the focus of attention as they moved restlessly between New York, Paris, and the French Riviera. From the time that they were married, Scott sold everything he wrote. He worked hard and played hard, and as his drinking increased, Zelda often completed an unfinished piece of his writing. . . . Their life together was one great, gaudy spree'.[1] Eventually their happiness was destroyed, especially by Zelda's schizophrenia, but their difficulties seem to have been due more to problems inherent in their separate personalities than to flaws in the relationship itself.

One way of extracting the most important information from the harmonic matrices is to concentrate only on:

(a) Aspects which are close (i.e. underlined in the matrix) and which involve one of the 'personal planets' (Sun, Moon, Mercury, Venus, Mars) of at least one of the partners. (Ascendant and M.C. can be added if the birth times are known to be accurate.)
(b) Aspects which involve the same 'personal' planet for both partners (Sun-Sun, Moon-Moon, Mercury-Mercury, Venus-Venus, or Mars-Mars), since these appear to be especially important in synastry.

Following these principles, the aspects which stand out in the matrices for Scott and Zelda are as follows:

Radical matrix

Firstly there is the Mars-Mars conjunction, which shows a strong link which is not so much sexual as concerned with exuberant activity. (Both Marses are in Gemini: Scott's is in the fifth house closely conjunct Neptune, and Zelda's is in the eleventh house closely conjunct Pluto).

Secondly there is the close sextile between Zelda's Mercury and Scott's Pluto. Zelda's Mercury is close to her Ascendant and so represents the face which she shows to the world; and the link with Scott's Pluto (in his fourth house) suggests that Scott finds in Zelda's personality a way to completely transform and revitalize his private life.

H4 matrix

Here the only aspect which qualifies is the Venus-Venus sextile. We have said that sextiles in H4 are to do with the tense and nervous search for pleasure. This, at least in public, was the nature of Scott and Zelda's relationship, and the Venus-Venus sextile epitomizes it perfectly.

H5 matrix

Here there are a number of close aspects, as we would expect in a relationship which displayed such tremendous style. Firstly there is the Mercury-Mercury conjunction, which is significant of their ability to communicate with each other in style and to write in the same style. This is backed up by the close square between Scott's Mercury and Zelda's Mars, which shows that their joint style was built out of a tense relationship between Scott's ideas and Zelda's actions: it was she who tried to bring his ideas to life, and act them out 'for real'.

The close trine between Scott's Mars and Zelda's Uranus derives a special importance from the fact that both of these planets are on the fifth cusp of their respective radical charts, and so are concerned with self-display. Scott wants to display himself as a man of action; Zelda wants to display herself as a person of unique and scintillating individuality. The trine in H5 shows that they were able to weld these characteristics together into a cheerful and vivid style of behaviour.

The close square between Scott's Venus and Zelda's Pluto may be an indication of some of the difficulties which eventually

SCOTT FITZGERALD

ZELDA FITZGERALD

H1	☉	☽	☿	♀	♂	♃	♄	♅	♆	♇	A	M
☉											☍	
☽								∠				
☿										*̲		
♀												
♂		∠			☌				☌	☌	∠	
♃												☌
♄						△̲						
♅				∠								
♆												
♇					☌				☌	☌		
A											☍	
M		☌				△						

H4	☉	☽	☿	♀	♂	♃	♄	♅	♆	♇	A	M
☉							*					
☽								☍		□		
☿										△		
♀				*								
♂		☍							☌			
♃							*					
♄						△						
♅				☍				□̳				
♆					⊻̲							
♇	*		(¯)									
A												
M												

H5	☉	☽	☿	♀	♂	♃	♄	♅	♆	♇	A	M
☉												
☽			☍									
☿			☌							*		
♀	*								□	△		
♂			□̳					∠				
♃			☍									
♄					∠	△						
♅					△̤				△			
♆								☌̲			☌	
♇				□̳		☌̲						
A												
M												

H7	☉	☽	☿	♀	♂	♃	♄	♅	♆	♇	A	M
☉												
☽	△̳											
☿										*		
♀						☌	☍				☌	
♂	☌			⊼								
♃		(☌)							∟̤	□		
♄												
♅		*						△				
♆	*̤											
♇		*						☌				
A			☍					☌				
M												

H9	☉	☽	☿	♀	♂	♃	♄	♅	♆	♇	A	M
☉		*			☌							
☽				△̳								
☿					△				△	☍		
♀			△̳									
♂												☌
♃		*			☌							
♄						☌						
♅										∟̤		
♆		△										
♇												
A												
M												

Fig. 3 Scott and Zelda Fitzgerald: Harmonic matrices

Scott Fitzgerald

	H1	*H4*	*H5*	*H7*	*H9*
☉	2 ♎ 19	9 ♈ 16	11 ♍ 35	16 ♎ 13	20 ♎ 51
☽	4 ♉ 28	17 ♌ 52	22 ♍ 20	1 ♐ 16	10 ♒ 12
☿	24 ♎ 49	9 ♋ 16	4 ♒ 05	23 ♓ 43	13 ♉ 21
♀	23 ♎ 14	2 ♋ 56	26 ♑ 10	12 ♓ 38	29 ♈ 06
♂	20 ♊ 37	22 ♒ 28	13 ♉ 05	24 ♎ 19	5 ♈ 33
♃	29 ♌ 20	27 ♏ 20	26 ♈ 40	25 ♒ 20	24 ♐ 00
♄	16 ♏ 12	4 ♎ 48	21 ♉ 00	23 ♌ 24	25 ♏ 48
♅	21 ♏ 49	27 ♎ 16	19 ♊ 05	2 ♎ 43	16 ♑ 21
♆	20 ♊ 19	21 ♒ 16	11 ♉ 35	22 ♎ 13	2 ♈ 51
♇	13 ♊ 43	24 ♑ 52	8 ♈ 35	6 ♍ 01	3 ♒ 27
A	4 ♒	16 ♌	20 ♊	28 ♒	6 ♏
M	29 ♏	26 ♏	25 ♋	23 ♏	21 ♓

Zelda Fitzgerald

	H1	*H4*	*H5*	*H7*	*H9*
☉	1 ♌ 04	4 ♌ 16	5 ♐ 20	7 ♌ 28	9 ♈ 36
☽	6 ♋ 35	26 ♈ 20	2 ♌ 55	16 ♒ 05	29 ♌ 15
☿	13 ♌ 49	25 ♍ 16	9 ♒ 05	6 ♏ 43	4 ♌ 21
♀	8 ♋ 07	2 ♉ 28	10 ♌ 35	26 ♒ 49	13 ♍ 03
♂	18 ♊ 53	15 ♒ 32	4 ♉ 25	12 ♎ 11	19 ♓ 57
♃	1 ♐ 03	4 ♐ 12	5 ♌ 15	7 ♐ 21	9 ♈ 27
♄	29 ♐ 39	28 ♓ 36	28 ♐ 15	27 ♊ 33	26 ♐ 51
♅	8 ♐ 42	4 ♑ 48	13 ♍ 30	0 ♒ 54	18 ♊ 18
♆	28 ♊ 00	22 ♓ 00	20 ♊ 00	16 ♐ 00	12 ♊ 00
♇	17 ♊ 11	8 ♒ 44	25 ♈ 55	0 ♎ 17	4 ♓ 39
A	8½ ♌	4 ♍	12½ ♑	29½ ♍	16½ ♊
M	1 ♉	4 ♌	5 ♍	7 ♏	9 ♑

Fig. 4 Scott and Zelda Fitzgerald: Radical and harmonic positions

overcame the relationship. Links between Venus and Pluto are to do with the search for dedication and single-mindedness, and squares in H5 can often have the character of a blind obsession. This Venus-Pluto link suggests that there was something in Scott's personality to which Zelda responded by pushing herself to the limit and sometimes beyond it, and this may have hastened the disintegration of her own personality.

H7 matrix
The close trine between Scott's Sun and Zelda's Moon is of course the perfect romantic link, suggesting that Scott and Zelda felt that they were perfectly suited for a life of happiness together; quite simply, they felt that they were in love. And the romanticism of this link is intensified by the close sextile between Scott's Sun and Zelda's Neptune, which suggests that Scott typified for Zelda a romantic ideal.

H9 matrix
The close trines between Scott's Mercury and Zelda's Venus, and between Zelda's Moon and Scott's Venus, show that each of them finds joy and pleasure in the other's Venusian qualities of charm, attractiveness and beauty, though in Zelda's case this is an instinctive heart-felt reaction while in Scott's case it is more intellectual. As in the case of Morecambe and Wise, the effects of these H9 links (which are essentially to do with joy and peace) may not be very obvious on the surface in a relationship which is so much concerned with activity and show, but at a deeper level they help to cement the relationship.

A difficult marriage

Our third case is that of a difficult marriage, which ended in divorce after many years of strain, tension and hostility between the partners. In this case I will not go through all the harmonic matrices, but will draw attention only to the H4 matrix, which is especially indicative of tension and difficulty in a partnership.

In the H4 matrix (Figure 5, p.212) there is a remarkable configuration in which his Venus and his Mars are both square to her Venus and her Mars. This would not have been possible, but for the fact that he has Venus semi-square Mars in his radical chart, and she has Venus square Mars in her radical chart. Thus the combination of Venus-Mars square Venus-Mars in the H4 matrix indicates that each of them heightens and aggravates the

HE

H4 (SHE)	☉	☽	☿	♀	♂	♃	♄	♅	♆	♇	A	M
☉	∟									△		
☽	☍	△					□					
☿										⚼		
♀				□	□				△			
♂				□	□				△			
♃												
♄		☌										
♅	*	☌								∟		
♆												☌
♇								☌				
A								☌				
M												

Fig. 5 'A difficult marriage': H4 matrix

other's difficulties in reconciling the masculine and feminine sides of their personalities, and this is likely to lead to great stress and tension, especially on the sexual plane. (Sun semi-square Sun in the H4 matrix will further aggravate these difficulties.) Thus, this is a case where it is only by referring to the harmonic matrices that we can discover the astrological causes of tension in the relationship. If we had looked only at the radical matrix, which contains three apparently very helpful trines, we would not have been able to predict that the relationship would be so difficult.

It seems, in fact, that squares (and possibly also semi-squares) in the H4 matrix are especially important in explaining disharmony in a relationship. To quote from Chapter 4:

> When two planets are square in the H4 chart, the person feels that they represent an obstacle which is preventing him from manifesting his qualities to the outside world. Often he will identify himself with one of the planets, and ascribe the qualities of the other planet to his perceived 'enemies' in the outside world. He may fight these 'enemies', and may provoke them into behaviour which makes it appropriate for him to fight; but he may also use their 'hostility' as a reason for inaction and for failing to develop the potentialities of the planet with which he identifies.

Therefore, if there is a square in the H4 matrix between planet A in one partner's chart and planet B in the other partner's chart, the first partner is likely to feel that the second partner (in the part of his behaviour which is represented by planet B) is hostile to that part of his own behaviour which is represented by planet A. Either he will fight against his partner through the medium of planet A, or else he will use the perceived hostility of his partner as an excuse for failing to develop the potentialities of planet A. The second partner will feel the same thing in reverse. Thus, in our example of the 'difficult marriage', each of the partners feels that the other partner's Venus-Mars nature is hostile to his/her own Venus-Mars nature: they may react to this by fighting, or by retreating into sullen inaction, or by oscillating between the two.

General hints on interpretation in harmonic synastry

If we draw up the harmonic matrices for any two people – even for two people whose mutual acquaintance is very slight, and is likely to remain so – we are likely to discover many harmonic links between their charts, and it would be easy to be carried away by this and to conclude that these two people are irresistibly drawn to each other and are capable of forming a very close relationship. But it should never be forgotten that a relationship is an inherently fragile thing, which can at any time be dissolved through non-astrological causes. Almost any two people, if forced to spend a long time together in a prison cell, would form a close relationship of one type or another, and the harmonic matrices can say a great deal about the nature of their relationship. But if two people meet casually, the effects of the harmonic links between their charts may never become apparent unless they choose to relate closely to each other and to reveal themselves to each other in all their complexity.

As in any other form of synastry it is important to study not only the links between the charts but also the separate charts of the two partners. Thus, in studying the H5 matrix, we should refer back not only to the partners' separate H5 charts but also to their radical charts, in order to discover what part each planet plays in the person's make-up. Particular attention should be paid to the following points:

(a) Planets which, in the radical chart of one of the partners, are especially indicative of his or her needs in relationships.

This includes especially planets in the seventh and eighth houses; it may include planets which rule these houses; planets in Libra; Venus in a man's chart and Mars in a woman's chart; and so on. For instance, if one partner has Mercury in the seventh house and the other partner has Jupiter in the seventh house, then a Mercury-Jupiter link in *any* of the harmonic matrices will be especially important.

(b) Links between identical planets (Sun-Sun, Moon-Moon, and so on). The harmonic in which such planets come together will indicate the way in which the partners will most easily be able to relate to each other with this side of their personalities. For instance, if the two Mercuries come together in H5, the partners will tend to communicate with each other in a stylized way; if they come together in H7, they will tend to communicate in an inspirational way.

(c) Very close aspects (underlined in the matrix). These will operate more powerfully and obsessively than aspects which are less exact.

(d) Planets in one partner's chart which are linked with a whole configuration of planets in the other partner's chart. For instance, the H5 matrix may show that partner A's Mercury is conjunct partner B's Moon, opposite his Venus, and square his Saturn. This indicates that partner B has a Moon-Venus-Saturn T-square in his H5 chart, which is of prime importance for understanding his personality. The connection with partner A's Mercury shows how partner A will respond to this pattern of planets in partner B's chart.

(e) In general, more attention should be paid to links shown in the top-left corner of the matrix (which shows links between Sun, Moon, Mercury, Venus and Mars) than to those in other parts of the matrix. Aspects involving the Angles should also be regarded as important *if* the birth times of both partners are known to be accurate.

The interpretation of specific aspects in harmonic synastry can well be based on the interpretations which have been suggested for the same aspects within an individual birth chart. For example, suppose that partner A has Venus square to partner B's Neptune in the H5 matrix. In Chapter 5 we said that when there is a square in H5 there will be 'expenditure of effort towards finding an appropriate style'; and in Chapter 10 it was suggested that Venus-Neptune links in H5 were concerned with 'finding stylized ways of expressing a universal love of humanity or sympathy for the human condition . . . developing an elusive charm, able to attract others by one's air of caringness and sensitivity'. Therefore a Venus-Neptune square between two partners in the H5 matrix may be to do with the effort towards jointly developing this kind of style: they will strive to develop a style of caringness and sensitivity in their dealings with each other, and perhaps also in their joint dealings with the outside world. One difficulty, however, may be to separate out the distinct roles and attitudes of the two partners. In this case it is partner A who supplies the Venus element (the search for charm and attractiveness), while partner B supplies the Neptune element (the selflessness and idealism). Thus, partner A finds partner B's selflessness attractive (and so seeks to attract it towards himself through his own charm), while partner B finds that partner A's charm fits in with his ideals (and so strives to respond selflessly to this aspect of partner A's behaviour). The exact way in which this works out will, however, depend on the sign- and house-positions of Venus and Neptune in the partners' radical charts, and on the other planets with which they are linked both in the radical charts and in the H5 charts.

Harmonic synastry, then, appears to offer a rich source of data for the analysis of relationships, and it is likely that further research and further experience by practising astrologers will greatly enrich our understanding of the ways in which the data can be interpreted and used in consultancy.

The harmonics of composite charts

Another approach to harmonic synastry is through the harmonics of composite charts. The composite chart of a relationship is the chart that results from taking the midpoint of (i.e. the half-way point between) A's Sun and B's Sun, the midpoint of A's Moon and B's Moon, and so on for all the other factors in the partners' charts, and entering them on a single chart. The technique has been fully described and discussed by Robert Hand.[2] (In one respect, however, a departure from Hand's technique is recommended. Hand suggests that, although the composite Midheaven is the midpoint of the two Midheavens of the partners' charts, the composite Ascendant – the cusp of the first house – should be calculated by taking the Ascendant corresponding to the composite Midheaven at the latitude where the couple are residing. This would seem to be an unnecessary complication – not least because it causes problems in the case of couples who, like Scott and Zelda Fitzgerald, are constantly on the move! I would prefer to regard the midpoint between the partners' Ascendants as the composite Ascendant and the cusp of the first house, and to calculate the cusps of the other houses in the same way. I believe that, since writing his book, Robert Hand himself has come round to this point of view.)

It is possible to draw aspects between planets in the composite chart using harmonic orbs, and also to calculate harmonic charts for the composite chart. Thus the H4, H5, H7 and H9 charts of the composite chart can be drawn in order to discover how the composite chart operates at each of the harmonic levels.

However, a technical problem arises when using harmonic orbs in a composite chart. The planets shown in the composite chart are not really planets, but midpoints between planets: thus the Sun is the midpoint between 'his' Sun and 'her' Sun, and so on. But in fact, for any midpoint on the circumference of a circle, the opposite point is equally valid. Thus, if 'his' Sun was at 0° Aries and 'her' Sun at 0° Leo, there would be two equally valid midpoints: 0° Gemini and 0° Sagittarius. It would be customary to draw the composite Sun at 0° Gemini as the 'nearer' of the two midpoints; but if there is another composite planet at 0° Sagittarius, it would really be conjunct with, and not opposite to, the composite Sun. Thus, in the composite chart, the conjunction and opposition are really the same; and from this it follows that the trine and sextile are really the same, since a planet which is trine to one of the two valid midpoints is sextile to the other. Therefore, we should have the same orb for the conjunction as for the opposition, and the same orb for the trine as for the sextile. Thus the recommended harmonic orbs for use in composite charts are as follows:

	Ordinary aspects	*'Close' aspects*
Conjunction / Opposition	6°	1°
Square	3°	30′
Trine / Sextile	2°	20′
Semi-square	1°30′	15′
Semi-sextile / Quincunx	1°	10′

These orbs apply to the radical composite chart and also to the H5, H7 and H9 composite charts. But they do *not* apply to the

H4 composite chart (or to any other even-numbered harmonic), since in an even-numbered harmonic chart the two opposite midpoints come together into a single point. Thus in the H4 composite chart we can use the same harmonic orbs as in an ordinary birth chart.

I have as yet done very little work with this technique, and I am putting it forward for detailed investigation by other astrologers. However, it seems likely that the harmonics of composite charts can be a useful supplement to the harmonic matrices which we have already discussed, since the information which they provide is essentially different. Whereas the matrices describe the attitudes and behaviour of the partners towards each other, the composite chart is more concerned with the *relationship as a thing in itself*.

As an example, we can take the H4 (p.218) and H7 (p.218) charts of the composite chart for Morecambe and Wise. Both of these charts are extremely strong. It is interesting that, whereas the H4 aspect-matrix for Morecambe and Wise contains relatively few aspects, the H4 composite chart (p.218) contains an unusually high number. The explanation would seem to be that in this case there is relatively little tension and strife *between* the partners (which would be indicated by 'hard' aspects in the H4 matrix), but that they are *jointly* engaged in a struggle against the outside world. Sun-Mercury-Venus opposite Pluto suggests their long and hard struggle (before they achieved fame and success) to dedicate themselves totally to their work and to maintain their Mercury-Venus charm in the teeth of all difficulties; and Mars exactly opposite Neptune (trine to Jupiter) suggests their efforts to escape from their everyday physical needs and to work towards a more long-term and idealistic goal, by means of exuberant action. The composite chart, and especially the fourth harmonic of the composite chart, is concerned with this kind of joint activity against the world. In fact, it seems likely that because the composite chart is a midpoint technique, the fourth harmonic is especially important in composite charts, as it is also in Ebertin's work[3] on the midpoints in an individual birth chart.

The seventh harmonic of Morecambe and Wise's composite chart is also extremely strong. Here the Sun-Mercury-Venus conjunction (carried over from the radical chart) comes together with both Mars and Uranus (it must be remembered that in an odd-numbered harmonic of a composite chart an opposition is identical with a conjunction), so that four of the five inner planets (Sun, Mercury, Venus and Mars) are joined together in

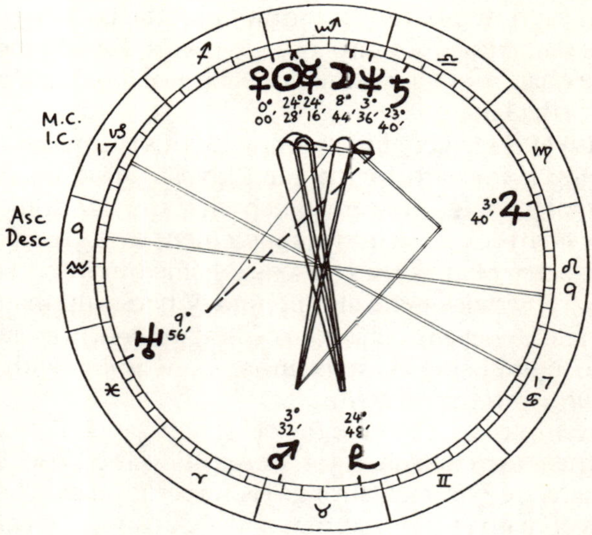

Morecambe and Wise: H4 composite chart

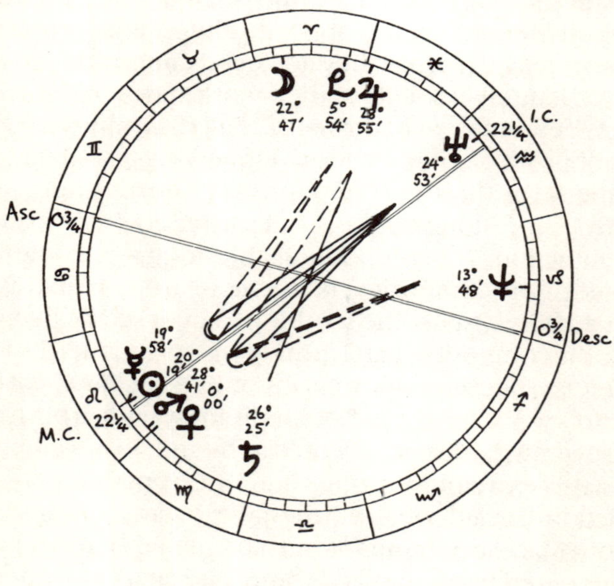

Morecambe and Wise: H7 composite chart

a single configuration with Uranus. This suggests that Morecambe and Wise are able – in an intuitive, inspirational, impulsive way (H7) – to make of their partnership a totally integrated 'thing', with a personality of its own, and with the quality of great originality (Uranus).

The ninth harmonic of Morecambe and Wise's composite chart is also quite strong; but the fifth-harmonic chart is unusually weak, containing virtually no aspects except for those carried over from the radical chart. This may seem surprising for a partnership which contains so much style.

Thus, it has to be admitted that our understanding of the harmonics of composite charts is at present rudimentary, and that further investigations will have to be made before they can be interpreted with confidence.

The 'relationship horoscope'

Another approach to synastry is the technique invented by Davison,[4] which he calls the 'relationship horoscope', but which has become known to some others as the 'Davison Composite'. The relationship horoscope is the chart for the moment in time which is exactly half-way between the birth times of the two partners, and for the point on the Earth's surface which is exactly half-way (in terms of latitude and longitude) between the birth places of the two partners. Thus, Wise was born at 3.00 a.m. GMT on 27 November 1925 in Leeds (53 N 48.1 W 34), and Morecambe was born at 11.00 a.m. GMT on 14 May 1926 in Morecambe (54 N 05.2 W 52): therefore, their relationship horoscope is the chart for 7.00 a.m. GMT on 19 February 1926 (the half-way date) in a village called Nappa (53 N 56.2 W 13), which is half-way as the crow flies between Leeds and Morecambe.

The relationship horoscope fulfils the same function for Davison as the composite chart does for Hand. Technically it is simpler than the composite chart. Unlike the composite chart, it is a real chart for a real time, and its planets are real planets and not midpoints: therefore, we can use the same harmonic orbs (in the radical chart and in the harmonic charts) as in an ordinary birth chart. Whereas in the composite chart oppositions are the same as conjunctions and sextiles the same as trines (except in an even-numbered harmonic), in the relationship horoscope all the aspects have their usual meanings.

However I have great doubts about whether the relationship

horoscope has any real validity. One of the arguments against it is the theoretical one: the planets shown in the relationship horoscope have no connection with the planets in the partners' charts, and it is difficult to see how they could be indicative of the relationship between them. Another argument is the practical one: if the composite chart is valid (and an increasingly large number of astrologers believe that it is), there is little justification for introducing another less well-tested technique whose purpose is identical with that of the composite chart.

If we take again the case of Morecambe and Wise, the relationship horoscope is a rather unremarkable chart, which (unlike the composite chart) does not give the impression that it is the chart of an exceptionally strong and original partnership. This is true especially of the radical chart and of the H4 and H7 charts. But in the fifth harmonic the story is different: the composite chart was weak in H5, but the relationship horoscope has a strong H5 chart in which Sun is closely trine to Uranus and square to Venus. This seems to be a good description of Morecambe and Wise's performing style: Sun trine Uranus showing a cheerful clarity and brilliance, and Sun square Venus showing the tenseness of their efforts to please their audience through their style. Venus conjunct Jupiter also seems to be an appropriate aspect. (But Mars conjunct Pluto, Mercury trine Saturn, and Venus trine Neptune are rather less appropriate, if the interpretations given in Chapter 10 for these aspects in the H5 chart can be applied to the relationship horoscope.)

Thus, this one case does not give a clear answer to the question of whether the composite chart or the relationship horoscope is the more valid technique. Readers are of course welcome to make their own investigations, setting up the radical and harmonic charts for each technique and deciding which technique gives a better description of the reality of a relationship. In my view, however, both the composite chart and the relationship horoscope should always take second place to the matrices of the aspects between the planets in the partners' charts.

CHAPTER 18
HARMONIC CHARTS FOR MUNDANE EVENTS

If harmonics are useful in the study of birth charts, they may also be useful in studying the charts of mundane events. I have not been able to do very much investigation in this area, but I have looked at a few cases which strongly suggest that the area is worthy of further study. Some of these cases will be presented in this chapter.

Volcanic eruptions

Krakatoa

The eruption of Krakatoa was the biggest bang in recorded history. Rocks were thrown thirty-four miles into the air, and the noise of the explosion was heard over one-thirteenth part of the surface of the earth. The island of Krakatoa disappeared, but tidal waves from it spread out to the neighbouring islands of Java and Sumatra and killed more than 30,000 people. Spectacular sunsets were seen all over the earth for several years because of the amount of dust thrown up by the explosion.[1]

The radical chart (p.222) for the Krakatoa eruption is remarkable chiefly for its shape, which gives a strong impression of 'everything going up in the air'. All ten planets are above the horizon and are grouped almost symmetrically around the M.C., with Jupiter close to the M.C. suggesting that the goal is expansion and freedom. We can see this chart from the point of view of the fire inside the earth, which chose this moment (M.C at 3° Leo) to break free from the confines of its environment.

But it is when we turn to the harmonic charts that we gain the clearest picture of how the *aspects* between the planets were appropriate to the occasion.

The H4 chart (p.223) shows Jupiter closely opposite Saturn (this was visible as a close semi-square in the radical chart), with both closely square to Mars. T-squares in the H4 chart are likely to result in a 'strong pattern of aggressive manifestation' (p.223). Jupiter opposite Saturn (with Jupiter close to the radical M.C.) suggests the urge to break free from restrictions, and Jupiter-Saturn square to Mars suggests that this will be achieved through forceful and direct action. Thus, this T-square is a strong indication of the type of event that was likely to happen at this moment.

The H5 chart (p.223) also contains a T-square (Moon-Neptune square to Saturn), but it is less powerful, though it gains strength from the position of Moon and Neptune on the M.C.-I.C. axis. But the H5 chart, like the radical chart, is remarkable chiefly for its shape, which shows the maximum possible dispersion of the planetary forces (ten planets in ten different signs). If the fifth harmonic stands for form and order, then this dispersion in the H5 chart represents the dissolution of form and order, and the creation of disorder and chaos (an impression that is reinforced by the emphasis on Moon and Neptune). This is of course highly appropriate to the Krakatoa explosion. All life in the vicinity of Krakatoa was destroyed, and it was many years before a new

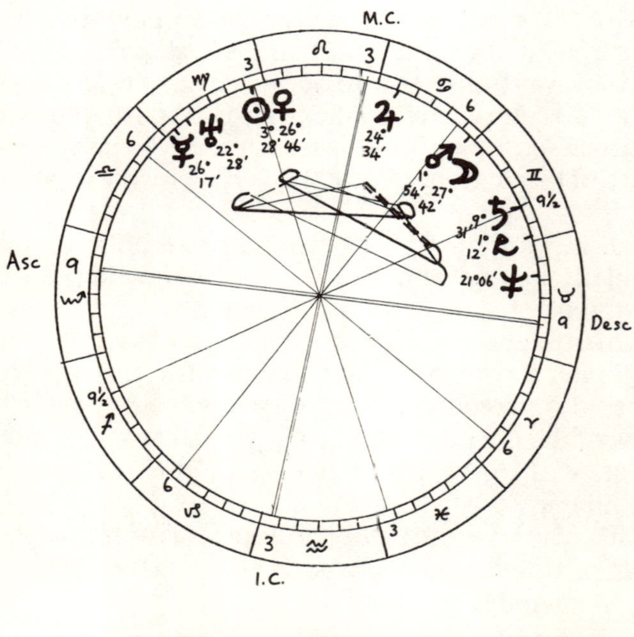

Krakatoa: Radical chart

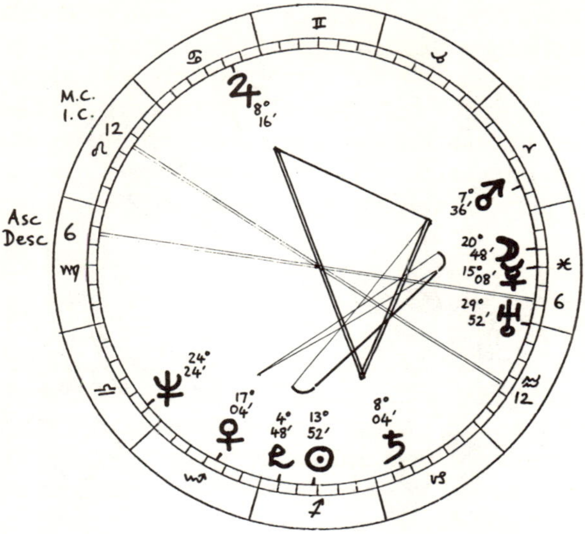

Krakatoa: H4 chart

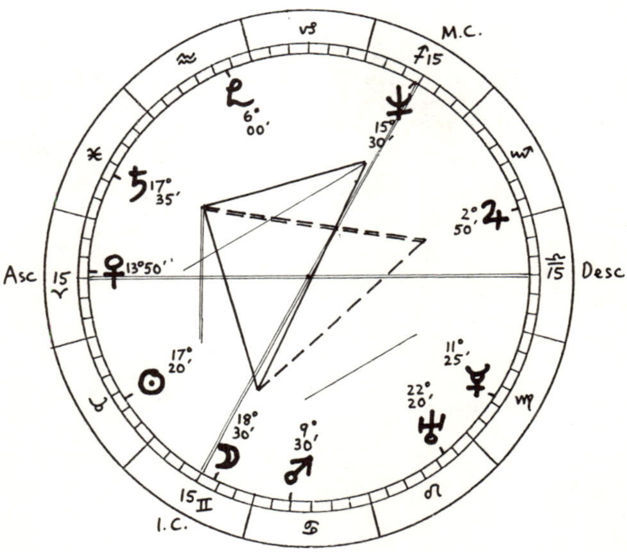

Krakatoa: H5 chart

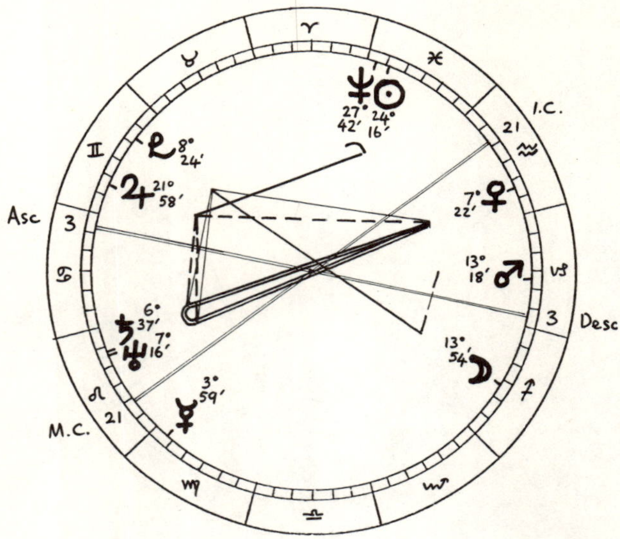

Krakatoa: H7 chart

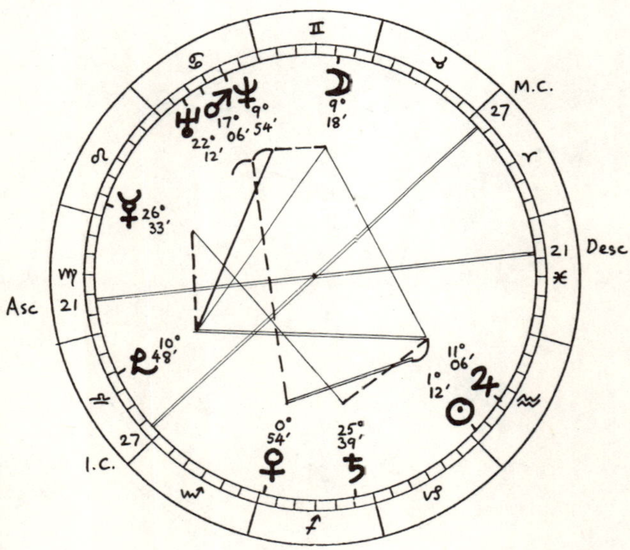

Krakatoa: H9 chart

island gradually rose from beneath the waves and life began gradually to return.

THe H7 chart (p.224) contains an extremely powerful (because extremely close) configuration of Saturn conjunct Uranus with both opposite to Venus. Saturn-Uranus aspects are notorious for accidents and sudden breakages, and this conjunction in H7 (the harmonic of inspiration and sudden impulse) suggests an inspired accident, an impulsive destructiveness. The opposition to Venus suggests that the impulse was to destroy the calm and peaceful beauty of the *status quo* and create a newer and more terrible beauty. Also, all three planets are closely linked to Jupiter near the Ascendant, suggesting again that this will be achieved by a magnificent display and by a process of breaking free. Thus, the H7 chart seems to describe the inspiration which moved the earth god or goddess at the time of the explosion.

The H9 chart (p.224) (as often with H9 charts) is more difficult to interpret, and perhaps we should not try too hard to do so since H9 represents joy and peace, and the eruption of Krakatoa was clearly not a time of joy and peace. But we can quote from the interpretations given in Chapter 10 for the three closest aspects in this chart: Jupiter trine Pluto, 'rejoicing in one's ability to achieve freedom through self-punishment'; Sun sextile Venus, (striving to) 'obtain joy and peace from appreciation of self'; and Mercury trine Saturn, 'rejoicing in being able to communicate effectively'. The earth goddess seems, in this chart, to be rejoicing in her own power, and in her ability to communicate clearly and unambiguously to the parasites who crawl around on her surface.

If we look at all the harmonic charts for Krakatoa together, we see that Saturn is the planet which links together the Mars-Jupiter-Saturn T-square in the H4 chart, the Moon-Neptune-Saturn T-square in the H5 chart, and the Venus-Uranus-Saturn pattern in the H7 chart. Thus Saturn, which in the radical chart is exactly on the eighth cusp, seems to be the main significator of the event; but we must look at all the harmonic charts in order to understand the complexity of the forces acting on Saturn at that moment.

Thus, the harmonic charts for Krakatoa seem to be very appropriate for a major volcanic explosion; and, since Krakatoa was the biggest explosion in history (or at least the biggest since the eruption of Santorini, whose time is unknown), we would expect it to exhibit these characteristics more strongly than any other case. Nevertheless, it would be unwise to build a theory on a single case. I have therefore looked at the charts of three other major eruptions: *Katmai* in Alaska, which was the biggest eruption of the twentieth century (although no one was killed because of the extreme remoteness of the area); *Mount Pelée* in Martinique, which was the most destructive of human life in the twentieth century (30,000 people killed instantaneously); and *Mount St Helens* in America, which, although not quite on the same scale as the others, was perhaps the most widely publicized and fully investigated volcanic eruption of all time. For reasons of space I am not presenting the harmonic charts for these eruptions in this book, but the planets' positions in the harmonic charts are presented in Figure 6 (below). For each harmonic in turn, we will summarize the extent to which the harmonic charts for these eruptions are similar to those for Krakatoa.

Katmai, Mount Pelée and Mount St Helens

Radical chart To a remarkable extent, the charts for these

Katmai

	H1	*H4*	*H5*	*H7*	*H9*
☉	11 ♊ 02	14 ♑ 08	25 ♓ 10	17 ♌ 14	9 ♑ 18
☽	3 ♑ 23	13 ♈ 32	16 ♑ 55	23 ♋ 41	0 ♒ 27
☿	23 ♉ 55	5 ♏ 40	29 ♐ 35	17 ♈ 25	5 ♌ 15
♀	1 ♊ 43	6 ♐ 52	8 ♒ 35	12 ♊ 01	15 ♎ 27
♂	2 ♌ 44	10 ♌ 56	13 ♐ 40	19 ♌ 08	24 ♈ 36
♃	10 ♐ 27	11 ♑ 48	22 ♍ 15	13 ♒ 09	4 ♋ 03
♄	25 ♉ 52	13 ♏ 28	9 ♑ 20	1 ♉ 04	22 ♌ 48
♅	3 ♒ 13	12 ♌ 52	16 ♊ 05	22 ♒ 31	28 ♎ 57
♆	21 ♋ 58	27 ♊ 52	19 ♎ 50	3 ♊ 46	17 ♑ 42
♇	28 ♊ 04	22 ♓ 16	20 ♊ 20	16 ♐ 28	12 ♊ 36
A	23½ ♍	4 ♓	27½ ♌	14½ ♌	1½ ♌
M	21 ♊	24 ♒	15 ♉	27 ♎	9 ♈

Mount Pelée

	H1	*H4*	*H5*	*H7*	*H9*
☉	16 ♉ 57	7 ♎ 48	24 ♏ 45	28 ♒ 39	2 ♊ 33
☽	24 ♉ 40	8 ♏ 40	3 ♑ 20	22 ♈ 40	12 ♌ 00
☿	27 ♉ 57	21 ♏ 48	19 ♑ 45	15 ♉ 39	11 ♍ 33
♀	1 ♈ 15	5 ♈ 00	6 ♈ 15	8 ♈ 45	11 ♈ 15
♂	8 ♉ 13	2 ♍ 52	11 ♎ 05	27 ♐ 31	13 ♓ 57
♃	15 ♒ 59	3 ♎ 56	19 ♌ 55	21 ♉ 53	23 ♒ 51
♄	27 ♑ 48	21 ♋ 12	19 ♉ 00	14 ♑ 36	10 ♍ 12
♅	20 ♐ 34	22 ♒ 16	12 ♏ 50	23 ♈ 58	5 ♎ 06
♆	29 ♊ 36	28 ♓ 24	28 ♊ 00	27 ♐ 12	26 ♊ 24
♇	17 ♊ 26	9 ♒ 44	27 ♈ 10	2 ♎ 02	6 ♓ 54
A	20 ♊	20 ♒	10 ♉	20 ♎	0 ♈
M	11½ ♓	16 ♑	27½ ♐	20½ ♏	13½ ♎

Mount St Helens

	H1	*H4*	*H5*	*H7*	*H9*
☉	27 ♉ 50	21 ♏ 20	19 ♑ 10	14 ♉ 50	10 ♍ 30
☽	21 ♋ 33	26 ♊ 12	17 ♎ 45	0 ♊ 51	13 ♑ 57
☿	4 ♊ 11	16 ♐ 44	20 ♒ 55	29 ♊ 17	7 ♏ 39
♀	1 ♋ 51	7 ♈ 24	9 ♋ 15	12 ♑ 57	16 ♋ 39
♂	4 ♍ 42	18 ♐ 48	23 ♉ 30	2 ♈ 54	12 ♒ 00
♃	1 ♍ 00	4 ♐ 00	5 ♉ 00	7 ♓ 70	9 ♑ 00
♄	20 ♍ 12	20 ♒ 48	11 ♌ 00	21 ♋ 24	1 ♋ 48
♅	23 ♏ 23	3 ♏ 32	26 ♊ 55	13 ♎ 41	0 ♒ 27
♆	21 ♐ 56	27 ♒ 44	19 ♏ 40	3 ♉ 32	17 ♎ 24
♇	19 ♎ 25	17 ♊ 40	7 ♑ 05	15 ♒ 55	24 ♓ 45
A	12½ ♋	20 ♉	2½ ♍	27½ ♓	22½ ♎
M	16 ♓	4 ♒	20 ♑	22 ♐	24 ♏

Fig. 6 Katmai, Mt Pelée, Mt St Helens: Radical and harmonic positions

three eruptions preserve the pattern of *symmetrical grouping around the M.C.* which was evident in the Krakatoa chart. In the case of Katmai, there are seven planets grouped around the M.C., with Pluto closest to the M.C.; two other planets (Moon and Jupiter) are symmetrically placed on either side of the I.C., but the pattern is slightly spoilt by the tenth planet (Uranus) which is also a long way below the horizon. In the case of Mount Pelée there is an almost symmetrical grouping around the M.C., with nine planets above the horizon and the tenth (Neptune) only slightly below it. In the case of Mount St Helens the planets are symmetrically grouped around the I.C. rather than the M.C.. But, since the M.C. is trine rather than square to the Ascendant, the chart gives the appearance of being tilted to one side: this is interesting since Mount St Helens erupted sideways. (It should be noted that the 'symmetry' is in terms of the area covered, rather than the number of planets. The number of planets on either side of the M.C. (or I.C.) may not be identical, but the outermost planets are always at approximately the same distance on either side of the M.C. or I.C.)

H4 chart All these H4 charts, like that for Krakatoa, show a very strong 'pattern of aggressive manifestation'. The H4 chart for Katmai has Mars-Uranus square to Saturn and semi-square to Neptune, and also Sun-Jupiter square to Moon: these two configurations are linked together by trines, sextiles and quincunxes. The H4 chart for Mount Pelée has Sun-Jupiter opposite Venus-Neptune and semi-square to Mercury-Uranus: also there is Moon square Pluto. The H4 chart for Mount St Helens has Mercury-Mars opposite Pluto and semi-square Uranus, also Sun square Saturn and semi-square Venus.

H5 chart All these eruptions have a pattern of great dispersion in H5, although not in such an extreme form as in the Krakatoa H5 chart. The dispersion is less marked for Katmai than for the others, which may be related to the fact that the Katmai eruption was less destructive of life.

H7 chart The Katmai eruption has an extremely strong H7 chart, creating (like that for Krakatoa) the impression of a 'special' chart for a remarkable event: Sun and Mars on the Ascendant are opposite to Jupiter and Uranus on the Descendant, and are also in a close Grand Trine with Mercury and Pluto; Mercury is with Saturn on the I.C., and Pluto is opposite Venus and semi-square Saturn. But this strength is not repeated in the H7 charts for the other eruptions. The H7 chart for Mount Pelée has Mercury-Jupiter on the I.C., and also Moon closely conjunct Uranus on the Descendant, and Mars closely conjunct Neptune; but there are no strong links between these configurations. Mount St Helens, like Katmai and Mount Pelée, has Mercury on the I.C. in H7, but the H7 chart for Mount St Helens is in general weak, with no strong links between planets.

H9 chart The H9 charts for these three eruptions have no obvious similarities either with each other or with the H9 chart for Krakatoa.

To a limited extent, then, there are similarities between the harmonic charts for these four eruptions (including Krakatoa). Perhaps the strongest similarity is the strong pattern of oppositions, squares and semi-squares in each of the H4 charts. But the planets involved are, to some extent, different in each case. There is no planet which is involved in the main H4 configuration for all four of these eruptions, although Mars, Jupiter and Uranus are each involved in three of the four cases.

More work will have to be done before we are able, from a study of harmonic charts, to predict when a volcanic explosion will occur. But such prediction would be very worthwhile. The charts which I have discussed are the charts for the 'big bang'; but the 'big bang' is usually preceded by a number of smaller explosions, and also by earthquakes, which warn people that there will soon be a major eruption. If an astrologer was able to predict the precise time of the 'big bang', this would provide evidence for astrology which might impress the scientists. Eventually, astrologers might be asked for advice on when to evacuate the area around a volcano.

Hiroshima

Volcanic eruptions are explosions over which man has no control. In contrast, we can look at the harmonic charts of a man-made explosion – the dropping of the atomic bomb on Hiroshima – which, although far smaller than the volcanic explosions we have discussed, seems more terrible because it had the express purpose of causing destruction and death.

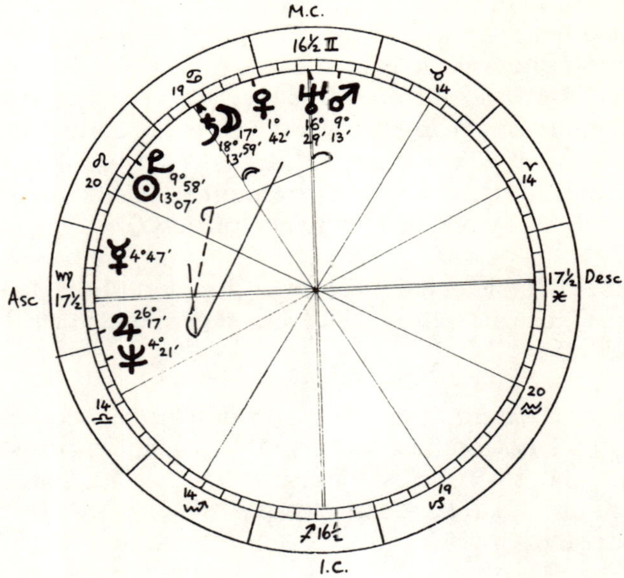

Hiroshima: Radical chart

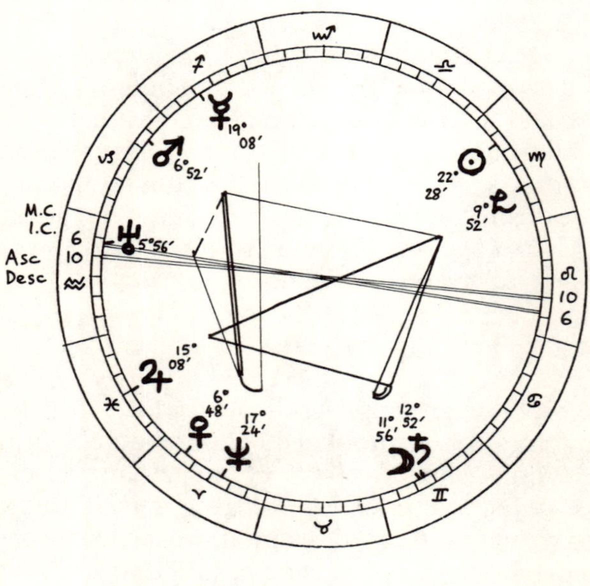

Hiroshima: H4 chart

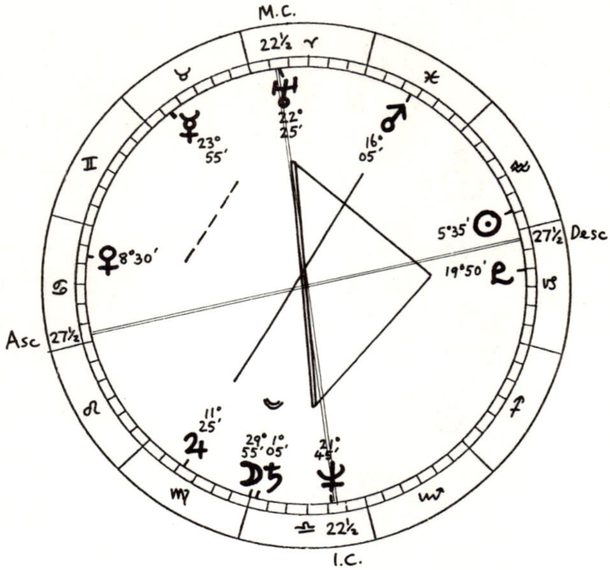

Hiroshima: H5 chart

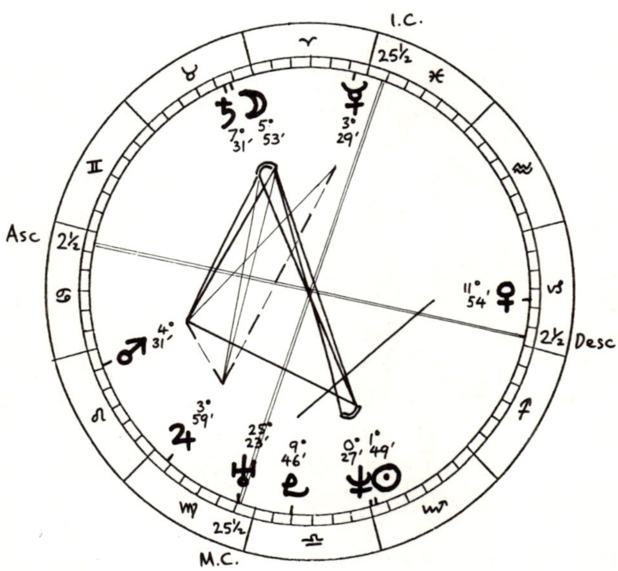

Hiroshima: H7 chart

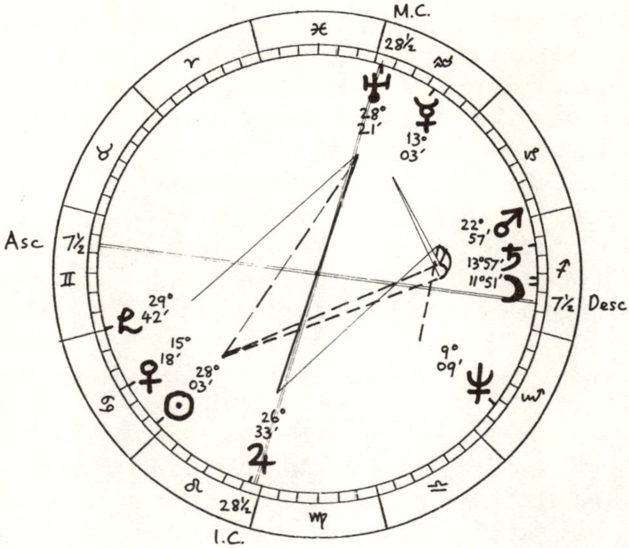

Hiroshima: H9 chart

Radical chart

The radical chart for Hiroshima (p.230), although remarkable for having Uranus exactly on the M.C. and Moon closely conjunct Saturn, contains few strong aspects; but (like the radical chart for Krakatoa) it seems to provide a graphic description of the event. It appears to tell the story from the point of view of the pilot who dropped the bomb. As he climbs into the air, he appears to be saying: 'I will leave behind me my ideals, my imagination, my religious principles (Jupiter and Neptune below the horizon); I will be ruthless, dedicated and single-minded (Sun conjunct Pluto); I will totally stifle and suppress my normal feelings and reactions, and concentrate solely on the task (Moon closely conjunct Saturn); I will look down on the beauty of what I am about to destroy (Venus); at the moment when I am directly over the city, I will drop upon it this "bolt from heaven" (Uranus on the M.C.); by doing this I will prove my manliness and courage (Mars).'

If it should seem fanciful to describe the pilot's feelings in this way, it should be remembered that the planets were actually looking down on him from these positions as he rose into the sky. Astrology is a description of reality.

H4 chart

In the H4 chart (p.230) the pattern of aspects becomes stronger. The Moon-Saturn conjunction is now square to Jupiter and Pluto, forming a T-square. The Moon-Pluto and Saturn-Pluto squares are what we would expect in connection with the dealing out of death and destruction; the involvement of Jupiter seems to imply that the deed has to be performed in a grand and spectacular way. But in addition we have Venus square Mars, which is the strongest aspect in the chart because it is almost exact. Squares in the H4 chart signify hostility, and so the H4 square between Venus and Mars (the feminine and masculine principles) is symbolic of rape: in this case, the rape of the passive Earth and its innocent inhabitants by the 'terrible thunder' descending from the hostile intruder in the sky. (If we look back at the radical chart, we see that Venus and Mars are on either side of Uranus which signifies the bomb itself: and, in the H4 chart, Uranus is linked to Venus and Mars by a sextile and a semi-sextile. The bomb is the instrument of rape.)

H5 chart (p.231)

Here Uranus is joined with Neptune and Pluto in a T-square, with Uranus closely opposite Neptune on the M.C.-I.C. axis. Since H5 signifies order and form, this T-square involving the three outermost planets is the clearest possible indication of the destruction of an old order and the creation of a new society with new forms and new values. Since Uranus, Neptune and Pluto move so slowly, this T-square is likely to have lasted for a considerable time; probably it signifies not merely the dropping of the bomb, but the whole process of the ending of the World War and the start of the post-war era, within which nuclear weapons (Uranus square Pluto) were to play an important part in the maintenance of order. But the opposition of Uranus and Neptune on the M.C.-I.C. axis indicates that the dropping of the bomb occurred at a climactic moment in this process.

H7 chart

In the H7 chart for Hiroshima (p.231), the radical Moon-Saturn conjunction is joined with Sun-Neptune and with Mars in a T-square. This chart seems to show how people were emotionally affected by the tragedy of Hiroshima: emotions caused by the suffering of the victims (Moon-Saturn) give rise to idealism and to a strong sense of purpose and identity with humanity (Sun-Neptune) and hence to a need for aggressive action (Mars). Thus we can see in this chart something of the original inspiration for the anti-nuclear movement. Perhaps the chart does not describe the terrible suffering of the actual victims of Hiroshima as clearly as one would have expected; but perhaps this suffering is best shown in the H4 chart (Moon-Saturn square Pluto).

H9 chart (p.232)
This is not a strong chart (Hiroshima was not a time of joy), and yet in the opposition between Uranus and Jupiter on the M.C.-I.C. axis we can sense something of the feeling that the dropping of the bomb will lead to peace, and to freedom from the hardships and struggles which the world had endured for six years.

The conquest of Everest

Finally we will look at the harmonic charts for a quite different type of event: the time when Hillary and Tensing became the first men to stand on top of the highest mountain in the world.

Radical chart (p.235)
Once again, the radical chart seems to provide a graphic description of the event. The explorers can be seen standing at the top of the mountain, having achieved their life's goal and the fulfilment of their dreams (Sun-Jupiter) as a result of rational planning and hard effort (Mercury-Mars). The Moon in a mundane chart often seems to represent the spectators, recipients or victims of an event, and in this case the Moon at the bottom of the chart seems to represent the rest of the world, looking up at the explorers on the mountain and admiring their Mercury-Mars qualities.

H4 chart (p.235)
The H4 chart provides a clearer framework than the radical chart for understanding the pattern of aspects. The meaning of the chart is probably more obvious if we turn it on its side, so that the M.C. (along with the Ascendant) is at the top of the chart.

Oppositions in the H4 chart lead usually to strong external manifestation. The H4 chart for Everest shows that the explorers were able to achieve their goal (Sun-Jupiter) only by strongly manifesting the qualities of Saturn-Neptune: stoic endurance coupled with idealism. Since Saturn and Neptune are the most selfless of the planets (see Chapter 9), the combination of the two

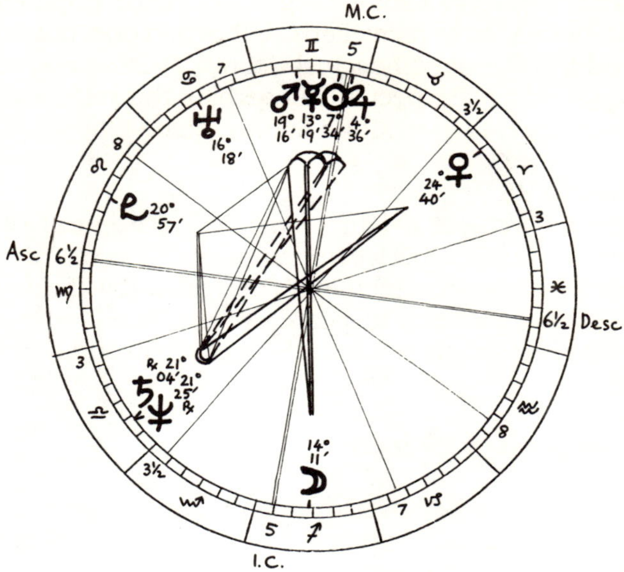

Everest: Radical chart

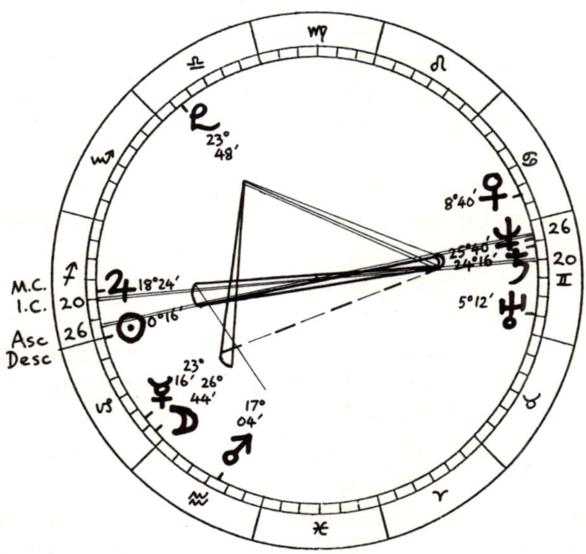

Everest: H4 chart

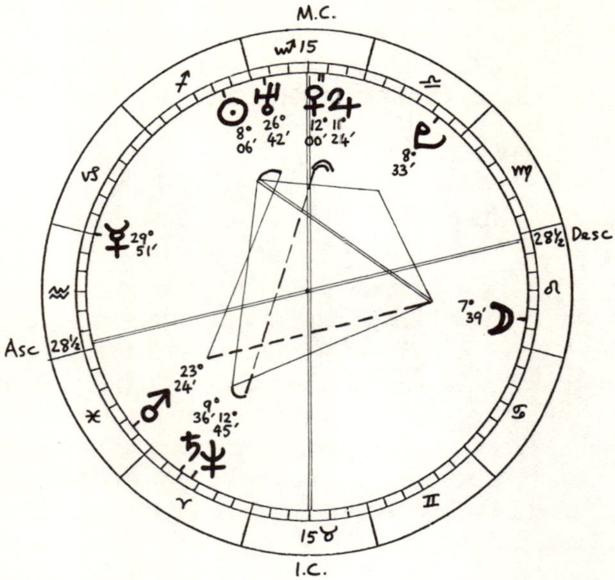

Everest: H9 chart

means that the climbers had to forget about their personal hardships in order to strive towards a higher goal; and yet the trines from Saturn-Neptune to Pluto show that this was mixed in with a quality of active self-castigation. But the close square between Mercury and Pluto shows that, even when they reached the top of the mountain, their problems were not resolved: this square seems to show them struggling to retain mental clarity in the face of the harshness of the environment, the lack of oxygen, and their own fatigue.

We will pass over the H5 and H7 charts, which contain no clear message (the H5 chart contains few aspects apart from Sun conjunct Pluto, and the H7 chart contains a large number of aspects but arranged in a complex and confusing pattern), and proceed straight to the H9 chart.

H9 chart (above)

The H9 chart contains a very strong kite-formation involving Sun, Moon, Saturn and Pluto. The close Sun-Moon trine in H9 clearly shows that the conquest of Everest was a time of great joy and ecstasy; and the links between this trine and the close Saturn-Pluto opposition show that this joy was achieved through deprivation and single-minded devotion to duty. Quite separate from this kite-formation is the close conjunction of Venus and Jupiter on the M.C.. I like to think that this represents the joy experienced by the climbers when standing on top of the world and surveying the beauty of the view which spread all round them.

Conclusion

I believe that these examples show that the use of harmonic charts is very worthwhile when studying the astrology of mundane events. Each harmonic chart seems to say something about the nature of the event at its own level. Thus, the H4 chart describes the type of effort which is being made, and the type of conflict which is taking place; the H5 chart describes the type of style which is being displayed or the type of order which is being created; the H7 chart describes the type of inspiration which is occurring, and the way in which people may be emotionally affected by the event; and the H9 chart shows the type of joy which is being experienced. But a strong harmonic chart will not *necessarily* lead to events of an appropriate type happening at that time; it probably only indicates a general readiness or proneness, which may or may not be taken up by people (in the case of man-made events) or by forces of nature (in the case of natural events). Also, the fact that the strong H7 charts of Krakatoa and Katmai do not recur in the cases of Mount Pelée and Mount St Helens provides a warning against assuming that we 'know the rules' before we have studied enough cases. Subject to these reservations, the study of the harmonics of mundane charts would seem to be well worth pursuing.

CHAPTER 19
HARMONIC TRANSITS

If harmonics can be applied to birth charts and to the charts of mundane events, we would expect them also to be applicable to predictive techniques such as transits and progressions. In this chapter we will look at how harmonics can be applied to transits, which (in my opinion) are by far the most important of the predictive techniques used by astrologers.

In the previous chapter we saw how, when the chart for a particular time is strong in a particular harmonic, an event may sometimes occur which partakes of the nature of that harmonic. For example, it seems that a volcanic explosion is more likely to be associated with a strong H4 chart (which denotes effort,

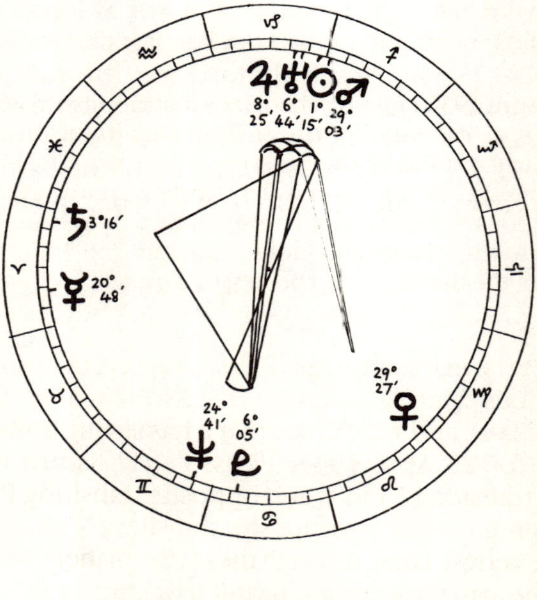

Noon GMT on 29 April 1937: H7 chart

action, conflict and tension) or a strong H7 chart (denoting impulsiveness, inspiration, wildness), rather than with a strong H5 chart (the creation of order, style, structure, form) or a strong H9 chart (joy, peace, happiness, enlightenment).

Let us look now at another example of a mundane harmonic chart. This is the H7 chart for noon GMT on 29 April 1937 (p.238) The Ascendant and M.C. have been omitted, and the Moon has also been omitted (since the Moon moves seven times as fast in an H7 chart as it does in a radical chart). Thus, this is not a chart for a particular moment or a particular place. It is an illustration of the H7 forces which were in operation throughout the day (although their strength will have been greater at some times of the day than at others, because of the movement of the faster planets).

We can see at once that this is a very strong chart. The close Venus-Mars trine seems to denote pleasure in physicality and sexuality; but the predominant theme is that of the assertion of individuality, freedom and self-fulfilment in the face of forces of destruction and restriction (Sun-Jupiter-Uranus opposite Pluto and square Saturn), with the close Uranus-Pluto opposition providing the most compelling force. Since this is an H7 chart, it is to do with inspiration: hence we would expect that this would be a good time for inspiration connected with these themes.

According to Roland Penrose in his biography of Picasso:

> On 29 April 1937 news reached Paris that German bombers in Franco's pay had wiped out the small market-town of Guernica, the ancient capital of the Basques. This gratuitous outrage, perpetrated at an hour when the streets were thronged with people, roused Picasso from melancholy to anger. Acting as a catalyst to the anxiety and indignation mingled within him, it gave him the theme he had been seeking. [1]

Thus it was on this date that Picasso received the inspiration for his famous painting 'Guernica'. Picasso's own H7 chart is also shown (p.240), and we can see that it has strong links with the H7 chart for 29 April 1937. Picasso's H7 Saturn is exactly conjunct transiting Uranus and opposite transiting Pluto, and his H7 Uranus is exactly opposite transiting Mars and sextile transiting Venus. Thus, through these two planets, Picasso was able to tune in to the inspiration which was 'in the air' at the time that the news of the Guernica massacre reached Paris. For Picasso

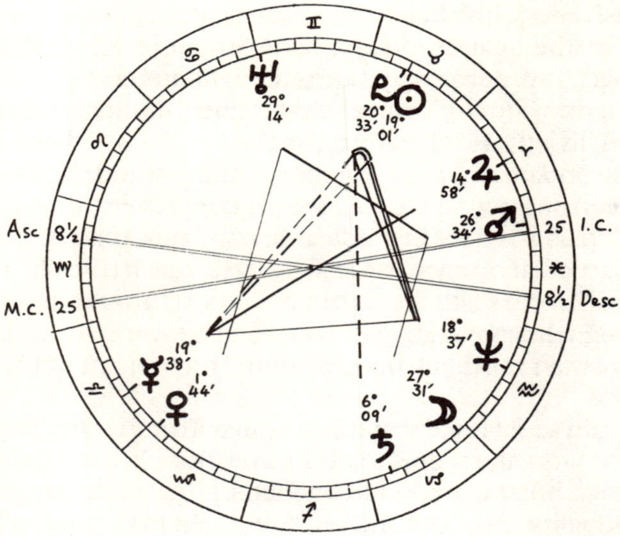

Pablo Picasso: H7 chart

it was primarily a Uranus-versus-Saturn issue – the cry of the individual against his fate – and this is the theme of 'Guernica', though we can also see in the painting the influence of the whole pattern of planets in the H7 chart of the day on which it was conceived.

It is important to stress that the inspiration was 'in the air', rather than being uniquely attached to Picasso. In fact, it could be argued that Picasso was rather poorly placed to take advantage of it; for, although his Saturn and Uranus are so strongly transited, there are no strong transits to the dominant configuration in his H7 chart, which is Sun-Pluto quincunx Mercury and square Neptune. But closer inspection shows that this is not the case: transiting Saturn is semi-square to Picasso's natal Sun and Neptune, and transiting Uranus and Pluto are semi-square to natal Pluto. The Sun-Pluto-Mercury-Neptune configuration represents Picasso's normal form of inspiration which is of a highly personal (Sun), self-revelatory (Pluto), rational and communicative (Mercury), imaginative and fantastical (Neptune) kind. On this occasion, the 'heavy' planets heavily aspecting his Saturn and Uranus drove Picasso into a form of inspiration which was at once more sombre and more electrifying than usual; yet at the same time he was motivated by the semi-squares to make full use of his normal sources of inspiration.

And yet there may have been many people whose natal planets were even more strongly touched by the transiting planets at that time than Picasso's were, and who may have felt even more strongly inspired and moved. But Picasso was the one who, because of his skill and experience as an artist and also because of his personal involvement in the tragedy of Spain, was able to convert his inspiration into a creative masterpiece.

Our second example is the H4 chart for noon GMT on 23 November 1963. This chart contains a very close configuration of Jupiter, Saturn and Uranus, with Saturn square to Uranus and quincunx to Jupiter; also, Mars is exactly semi-square to Saturn and Uranus (though, because of the greater speed with which Mars moves, this last aspect did not prevail throughout the day). Conflicts between Saturn and Jupiter-Uranus have the air of 'bringing down the mighty from their seat', or of shooting down something which is flying high and rejoicing in its freedom; they could therefore correlate with murder, or at least with the kind of murder which is motivated by envy. Thus we could say that murder was 'in the air'; and in fact the previous day (22 November, on which the H4 link between Saturn, Jupiter and Uranus was even closer and on which Mercury in H4 was conjunct Uranus, square to Saturn, and trine to Jupiter) was the date of the assassination of President Kennedy. 23 November, however, was the date of the first child-murder[2] committed by the sadistic Moors murderer, Ian Brady, together with his accomplice Myra Hindley.

If we compare the H4 chart for 23 November 1963 (p.242) with Brady's own H4 chart (p.242), we see that the links between them (or at least those which came to a head on this particular day) are entirely concerned with Brady's Uranus and Neptune. Brady's H4 Uranus is trine to transiting Jupiter and Uranus and semi-sextile Saturn, and his H4 Neptune is conjunct transiting Mars and semi-square to transiting Saturn and Uranus. Brady, therefore, was motivated on this day by Uranus-Neptune drives: the urge to do something which was on the one hand original, unique and thrilling, and on the other hand symbolic of a wider ideal or fantasy. Because we are dealing with the H4 chart, we are dealing with a drive towards *action* (and especially aggressive action against the outside world) rather than towards inspiration. And yet we must stress that the origins of Brady's sadistic fantasies

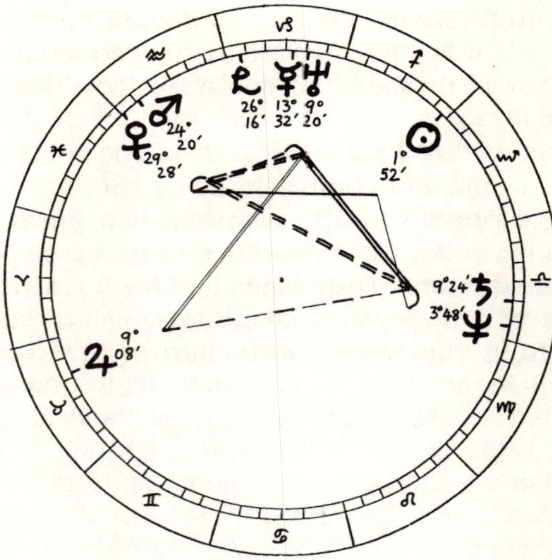

Noon GMT on 23 November 1963: H4 chart

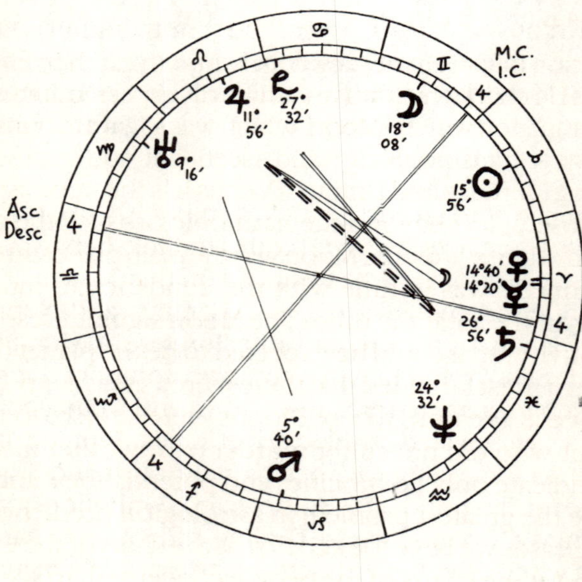

Ian Brady: H4 chart

lie not in his H4 chart but in his H7 chart, in which there is a close conjunction of Mercury, Venus, Neptune and Pluto (all trine to Saturn). Brady was often inspired to think about murder, but it was on this date that he felt compelled to convert his thoughts into action.

We can now look at the situation from the point of view of the murder victim. The chart of Brady's victim is not available, but President John F. Kennedy, as we have said, was assassinated on the previous day (22 November 1963, 12.30 p.m. CST, Dallas). The positions of the planets and Angles in the H4 chart for the time of Kennedy's assassination (which are very similar to the positions shown in the noon chart for 23 November) are as follows:

☉28♏56	☽15♍00	☿ 9♑00	♀25♒48
♂22♒08	♃ 9♉16	♄ 9♎16	♅ 9♑16
♆ 3♎44	♇26♑16	Asc. 8 ♎	Mc. 16 ♐

The positions in Kennedy's own H4 chart are as follows:

☉ 1♑20	☽ 8♒52	☿22♎24	♀ 7♒00
♂13♎48	♃ 2♏12	♄18♑40	♅ 4♏52
♆10♌44	♇13♈04	Asc. 20♊	Mc. 4 ♋

A comparison between these two lists shows that the planet in Kennedy's H4 chart which was most heavily aspected at the time of the assassination was the Moon, which was square to transiting Jupiter, trine to transiting Saturn and Ascendant, and semi-sextile to transiting Mercury and Uranus. (Venus has the same aspects but less closely; it is beyond the harmonic orbs for the semi-sextiles.) In other words, the Moon was the main planet through which Kennedy was in tune with the 'tendency to murder' which was in the air at the time. The Moon signifies response and receptivity, so it could be expected to be the planet of the victim. But it must be added that there were a great many other people alive at the time whose Moons were subject to these transits but who did not suffer murder or any other mishap. Transits indicate only possibilities and potentialities, and this is I believe the greatest problem in astrological prediction.

In these cases we have looked *first* at the chart of the event, and *second* at the charts of the people concerned, to see how they were connected to the chart of the event. This is the reverse of the usual astrological procedure, which is to start with the chart of the person and then to look at the transiting planets as though they 'belonged' in some way to the person. But I believe that the procedure which we have followed corresponds more closely to the way that transits actually work. The chart for a particular time may create a proneness to a particular type of event, and the harmonic charts may allow us to study this proneness in more detail: the H4 chart may indicate a proneness to action and conflict, the H5 chart may show a tendency to develop a type of order or style, the H7 chart may show a tendency towards a type of inspiration, and the H9 chart may show a tendency towards experiencing peace and enlightenment. When we have obtained some idea of the nature of this proneness, we can look at the charts of particular people to see to what extent, and in what way, they are likely to be affected by it.

But the procedure of starting with the chart for the event is extremely difficult in ordinary astrological prediction. If we are trying to predict the course of a person's life over a period of time (say a year), then we clearly do not have time to construct a separate chart – let alone a whole series of harmonic charts – for each day during that period, so that we can discover what types of 'proneness' exist on each day and then relate these to the chart of the client. Eventually, computers may provide an answer to this problem. It may already be possible to programme a computer to provide a moving visual display, showing how the aspect-patterns in the chart (that is, the chart for the time for which predictions are being made) change over time, and pinpointing the times when these patterns reach maximum strength and also the times when they come into closest contact with the natal planets in the client's chart. Such a programme could be developed both for the radical chart and for each of the harmonic charts, and could be an extremely powerful tool in astrological prediction. (I would like to extend an open invitation to any computer expert to help in developing such a programme.)

Table of harmonic transits

The astrologer who does not have access to this type of computer programme, and who has to calculate transits by hand, will have to resort to a compromise procedure, which will not reveal the full complexity of harmonic transits but will at least show some of the most important features. The procedure outlined below is a simple adaptation of the normal table of transits, as taught for instance by the Faculty of Astrological Studies. We can call our own method the table of harmonic transits.

In the traditional table of transits, the following aspects are shown between transiting and natal planets: conjunctions, oppositions, trines, squares, sextiles, semi-squares, semi-sextiles and quincunxes. In effect this means that conjunctions are shown in the following harmonics: H1, H2, H3, H4, H6, H8 and H12. In the table of harmonic transits, on the other hand, conjunctions are shown in the first nine harmonics: H1, H2, H3, H4, H5, H6, H7, H8 and H9. The decision to stop at H9 is of course an arbitrary one, and a good case can be made for the inclusion of H10 and H12 (and maybe also of other harmonics such as H14 and H15). But I have found in practice that, with the inclusion of nine harmonics only, one has as much information as the mind can cope with. (As so often in astrology, the problem is that of 'seeing the wood for the trees'. The inclusion of further details may provide additional insights, but it may also hinder one from seeing the overall pattern.)

♈0°21′	☉ 8
0°46′	☽ 3
3°15′	☿ 9
7°	Mc18
8°36′	♃ 5
8°58′	♆ 7
9°58′	♂ 7
10°	Mc7
11°14′	♀ 9
11½°	Asc18
13°19′	♇ 7
14°05′	☿ 3
15°	Mc10
↓	↓

Fig. 7 List of harmonic conjunctions (extract)

A necessary first step towards the table of harmonic transits is to compile a list of harmonic conjunctions for the planets and angles in the birth chart: that is, a list of zodiacal positions at which a transiting planet is in conjunction with any of the natal planets in any of the first nine harmonics. An extract from a list of harmonic conjunctions is shown in Figure 7 (above): this exact covers only the first half of Aries, but the full list extends right round the zodiac. The extract shows that a transiting planet at 0°21′ Aries is conjunct the natal Sun in H8 (that is, semi-square to the Sun in the radical chart); a planet at 0°46′ Aries is conjunct the Moon in H3, and a planet at 3°15′ Aries is conjunct Neptune in H9. Only the first nine harmonics are shown, except in the case of the Ascendant and M.C. for which H10, H14 and H18 are also included. This is because a planet which is conjunct the Ascendant in H10 is also conjunct the Descendant in H5, and the Descendant (and I.C.) are regarded as equally important as the Ascendant (and M.C.). (But we may decide to omit the Ascendant and M.C. from the list, especially for the higher harmonics, if the time of birth is not accurately known.)

The construction by hand of the list of harmonic conjunctions is not difficult, but is lengthy and tedious. The traditional aspects can be calculated (or observed from the chart) in the usual way. For quintiles (conjunctions in H5), start from the natal position of each planet, add on 72° (or 2 signs plus 12°), and continue to do so until returning to the natal position. For noviles (conjunctions in H9) add on 40° (or 1 sign plus 10°) in the same way. For septiles (conjunctions in H7) add on 51° 25½ minutes, and there is no escape from some very repetitive addition. However, the list of harmonic conjunctions could easily be calculated by a simple computer programme, and could be made available at low cost by the computer organizations which calculate birth charts.

Once the list of harmonic conjunctions has been constructed for a birth chart, it is a simple matter to compile a table of harmonic transits for any year in the life. Thus, for the case given in Figure 7, if a planet is transiting through the first half of Aries, enter ☉ .8 for the date on which it reaches Aries 0°21, ☽ 3 for the date on which it reaches Aries 0°46′, and so on.

An extract from a table of harmonic transits for a particular year is shown in Figure 8 (p.246). The following notes will help to clarify the contents:

1 *Columns 1-5* These columns contain the harmonic transits formed by the outer planets Pluto, Neptune, Uranus, Saturn and Jupiter. The house occupied by the transiting planet at the start of the year is shown at the head of the column. If the planet moves into a different house during the year, the new house is entered in brackets in the appropriate month. (In the example shown, Neptune moves into the tenth house

1982	Col. 1 ♇ 7th	Col. 2 ♆ 9th	Col. 3 ♅ 8th (☽1)	Col. 4 ♄ 7th	Col. 5 ♃ 7th (♂1) (♅2)	Col. 6 ♂ ♀ ☿ ☉	Col. 7 NM	Col. 8 ☽p. 6th	Col. 9 (♂p:♄p4) (☉p:♇4)
JAN	St→Mc6	28:♆7	11:♀5 12:♆9 17: Asc 10		1:♀8 3: Asc 8 19:♄3 26:♇7		25♒4°54′ 11th	(29th) ♌ 23°26′	☉p:☉8
FEB		10:♄5				11:♀ St:♄8 13:☿ St. Mc 10	23♓4°56′ 12th	♂5 ♌ 24°26′	
MAR	23:♇4 27:☿5	Mc1 (10th) St→♂7		14:☽9	25:♇7 31:♄3	12:☉: Asc 1 19:☿:♄1 27:☿ : Asc 1	25♈4°26′ 1st	♄9 ♌ 25°25′	
APR		Mc1 (9th)			17: Asc 8 19:♀8 24:♂1 25:♆2, Mc 7 26:☉9	14:♀ :♄1 18:☿ :♅1 26:☉ :♅1 26:♀ : Asc 1	23♉ 3°21′ 1st	Asc 7 ♌ 26°24′	
MAY		17:♄5 31:♆7	2: Asc 10 8:♆9 9:♀5	5: Asc 7 20:♇9 31:☽8	2:♆7 6:☿9 31: Asc 18	12:♂St:☉8 21:☿ St:(☉6)	23♊1°44′ 2nd ☽2 Mc 7 (♂7) (☿5)	Mc 3 ☿p. 6 ♌ 27°23′	
JUN			29:♅7 (☽1)	13:♅9 St☉6 22:♅9 (☽8)	4:♀9	4:♀ :♅1 18:☉: Mc 2	21♊29°47′ 4th ♄8	♌ 28°22′	(☉p:☽6) (☿p: Mc 6)

Fig. 8 Table of harmonic transits (extract)

in March when it crosses the M.C., and moves back into the ninth house in April when it recrosses the M.C.).

2 Within these columns, transits are shown for the date on which they become exact. Thus, for example, transiting Neptune was exactly septile to natal Neptune on 28 January.

3 If a transiting planet becomes stationary *within orbs of* an aspect to a natal planet, this is entered on the table, as in the cases of Pluto stationary close to MC 6 in January, and Neptune stationary close to ♂ 7 in March. The orbs used for this purpose (following the principle of harmonic orbs) are as follows:

(conjunction in:)	H1	2°00′
	H2	1°00′
	H3	40′
	H4	30′
	H5	24′
	H6	20′
	H7	17′
	H8	15′
	H9	13′

4 If an aspect remains within orbs (using the same orbs as in Note 3) for a long period, we may enter lines which show its duration. Thus, in the example given, the aspect ♇ : ♇ 4 lasted for the whole of March and the first part of April, and the aspect ♆ : MC 1 lasted for the whole of the six months shown in the table.

5 *Column 6* This column shows aspects formed by transiting Mars, Venus, Mercury and Sun. Normally, only conjunctions are shown (and oppositions to the Ascendant and M.C.). But, if Mars, Venus or Mercury becomes stationary within orbs of a harmonic aspect to a natal planet (using the same orbs as in Note 3), this is entered on the table (as in the case of ♀ st: ♄ 8 in February).

6 *Column 7* This column shows the New Moons. Against each month is shown the date of the New Moon, its zodiacal position, and its house-position in the birth chart. The date is ringed if the New Moon is an eclipse. If the New Moon is within orbs of a harmonic aspect to a natal planet (using the same orbs as in Note 3), this is entered in the table. (In the example given, under May, I have entered ♂ 7 and ☿ 5 in brackets, as they were just outside the permitted orbs.)

7 *Columns 8 and 9* These columns (to the right of the thick line) contain secondary progressions, rather than transits. I have an open mind about the validity of secondary progressions, but I have included them in the table in order to show that harmonic progressions can be derived from the list of harmonic conjunctions as easily as harmonic transits. (Any other directional method, such as solar arc, one-degree, or converse progressions could equally well be used.)

8 Column 8 contains harmonic aspects formed by the progressed Moon. Column 9 contains harmonic aspects formed by progressed Sun, Mercury, Venus and Mars: typically these last over a long period, so lines are drawn showing the period during which the aspect is within orbs (using the same orbs as in Note 3). In the example shown, ⊙ p: ⊙ 8 became exact in January, but ⊙ p: ♄ 4 and ♂ p: ♄ 4 became exact before the period shown in the table, and ⊙p: ☽6 and ☿ p:MC6 became exact after the period shown in the table.

Figure 8 shows an extract from a table of harmonic transits for a whole year; but, if we wish to do very detailed predictions for a particular month, it is possible to supplement this with a table showing the day-to-day harmonic aspects formed by transiting Sun, Mercury, Venus and Mars during that month. We can call this a table of harmonic mini-transits, and an extract from such a table is shown in Figure 9 (p.250). This supplements the information on March 1982 given in the main table of harmonic transits (Figure 8, p.247).

Prediction is not my main interest in astrology, and as yet I have compiled a list of harmonic conjunctions for only one individual (myself) and have constructed a table of harmonic transits for only two years and a table of harmonic mini-transits for only four months. Moreover, I have not been consistent in studying the tables to see whether they provide information which corresponds to the reality of developments in my life. In so far as I have studied the tables, however, I do feel that they often provide valid information, and are at least an improvement on the traditional tables of transits which contain no quintiles, septiles or noviles. In particular I feel that (using the interpretations for particular aspects which were proposed in Chapter 10) they provide valid information about my *moods* (and moods are of course an important determinant of actions and of the events in one's life).

MARCH 1982

	♂ T 7th ℞	♀ T 11th	☿ T 12th	☉ T 12th
1			☿ 2 ♃ 4	☉7, ♇8
2		Asc 7	♀5, ☉2	
3		☽ 6	Asc 10, ♅9 ♇ 9	♆2
4			♆ 7	☽ 7
5			Mc 7, ♂ 7	Mc 14, ♅ 7 ♃ 3
6		Mc 10		
7			☽ 9 ♇ 7	Mc 9
8			♆ 9	
9			♅ 5	
10		♃ 9 ♅ 4	♀ 7, Asc 14	☿ 5
11	Asc 7	♂ 4 ♀ 8 (12th)	Mc 6	♅ 8, ♂ 8
12		Asc 8 Mc 9	♃ 7	♀ 2, ☉5 Asc 1 (1st)
13		♆ 5		
14	♇ 9		☽ 4 ♀ 9, Asc 18	
15	☽ 8		♇ 5	

Fig. 9 Table of harmonic mini-transits (extract)

Really, however, I am not reporting on the success or failure of this method of prediction, but merely suggesting it as a method by which astrologers may be able to make some use of harmonics in their predictive work without radically altering their techniques. Readers will have to judge from their own experience whether the insights gained from the table of harmonic transits are worth the effort of constructing it. I believe that it is likely to be an improvement on existing methods, but will still fall a long way short of the ideal: for the table of harmonic transits shows only *conjunctions* in the first nine harmonics, whereas the examples which we gave of Picasso and Brady at the beginning of this chapter show that major events are often dependent on complex patterns of aspects within particular harmonic charts. In fact, the more I venture into this area, the more I feel that, although transits *are* valid, the forces on which they depend are so complex that the astrologer who attempts prediction (without the benefit of clairvoyant powers) does so at his own peril, and perhaps also at the peril of his clients. The study of past events (such as the eruption of Krakatoa or the painting of 'Guernica') is fascinating, but the prediction of future events is hazardous in the extreme. And anyway, why do we want to predict the future? I agree with Dianne Binnington when she says:

> If we really wish to help and to heal, we should concentrate solely on the natal chart, for without the self-knowledge that this affords, there can be no 'presentness' and, therefore, no peace . . . . Why offer information about the future when it obviously cannot be used wisely without self-knowledge and does nothing to realize and heal the inner state? . . . The snare of prediction is that it gives us to think that with fore-knowledge we have some control over our destiny; but it ignores the present and, in so doing, robs us of our real autonomy.[3]

Harmonic directions

I will conclude this chapter by referring briefly to the technique of prediction which has become known as 'harmonic directions'. This technique is introduced by John Addey in *Harmonics in Astrology*.[4]

In this technique, each harmonic of the birth chart represents a year of the life. Thus, by studying a person's forty-eighth harmonic chart, we can obtain information about likely developments in his life in his forty-eighth year, or about the year in which he reaches the age of forty-eight. The harmonic chart for the year is said to set the general tone for the year. We may not be able to predict specific events from it, but we may be able to identify the major themes which preoccupy the person during the year in question. In short, the harmonic chart for the year performs the same function as that performed by the solar return chart in traditional astrology.

I am very doubtful about this technique, for the following reasons:

1 It does not seem to work well when applied to past years of my own life; in fact, it works less well than the solar return

chart. (This argument is not, of course, conclusive in itself.)

2 Logically we would expect the first-harmonic chart (i.e. the radical chart itself) to apply to the first year of life, the second-harmonic chart to the second year of life, and so on; hence the forty-eighth harmonic chart would refer to the forty-eighth year of life, which is the year *before* one's forty-eighth birthday. But in harmonic directions the whole process is moved one year later, so that the radical chart relates to the second year of life, the second-harmonic chart relates to the third year of life, and the forty-eighth harmonic chart relates to the year *after* the forty-eighth birthday, which is in fact the forty-ninth year. This seems extremely illogical. Perhaps we do not have a right to except the cosmos to be logical (although on the other hand perhaps we do, as I shall argue in Chapter 23), but the fact remains that in other respects it does seem to display a sublime logicality, so that this sudden outburst of illogicality gives rise to severe doubts.

3 The interpretation of harmonic directions depends heavily on aspectual links *between* the harmonic chart and the radical chart. For example, if Uranus in the forty-eighth-harmonic chart is conjunct to Mercury in the radical chart, then this is held to lead to events of a Mercury-Uranus nature during the forty-eighth (or rather forty-ninth) year. But (for reasons which I will discuss further in Chapter 21) it is very doubtful whether it is admissible to use aspects *between* the harmonic chart and the radical chart. Uranus in the harmonic chart may appear to be in Aries (when its radical position is, for instance, in Gemini), but it is doubtful whether it really is in Aries in any meaningful sense and, if it is not in Aries, it cannot be said to be in conjunction with radical Mercury in Aries.

4 Another argument against the use of harmonic directions is that even if the technique is in theory valid, it is impossible to use it for events in adult life unless the time of birth is *very* accurately known. Any inaccuracy in the radical chart will be multiplied *forty-eight times* in the forty-eighth harmonic so that, for instance, an error of 1¼ degrees in the Midheaven, caused by an error of five minutes in the birth time, will cause the Midheaven in the H4 chart to be 60 degrees (or two whole signs) away from its true position. (Even the Moon will be 2 degrees away from its true position if there is an error of five minutes in the birth time.) This is not necessarily an insoluble problem: John Addey and

Charles Harvey have recommended that events in early childhood should be used to rectify the chart so that the higher harmonics can be used with greater confidence. But such information about early childhood is not always available; and in any case, it may seem unwise to use a technique which relies so much on rectification.

5 Underlying these arguments (and especially arguments 2 and 3) is a more profound question about the nature of harmonics. At the beginning of this book I said that the harmonic chart was only a device which enabled us to see more clearly a part of the aspect-pattern in a birth chart; and all the uses which I have suggested for harmonic charts are consistent with this. But in harmonic directions, the harmonic chart is elevated to the status of a thing-in-itself, a chart in its own right, which can be *related to* the birth chart as though it existed separately from it. Moreover, in their other uses, harmonic charts are concerned with *fractional* numbers (for instance, the H5 chart is concerned with *fifths* of the circle), but in harmonic directions they are concerned with *ordinal* numbers (numbers in a series): for instance, the fifth year of life is not one-fifth (or two-fifths) of anything, but is number 5 in a series of indeterminate length. Thus the harmonic charts are being put to a use which is *entirely different* from their other uses, and which (unlike all the other uses which we have suggested) involves a radical change in our conception of how astrology (and therefore the cosmos) works. Of course it is not impossible that harmonic directions are valid; but I for one cannot accept such a concept (which would involve rethinking all my other assumptions about astrology) unless there is unshakable evidence for their validity, and at present this evidence does not exist.

For all these reasons I do not feel able to recommend the use of harmonic directions until further research has shown the method to be valid. But readers are of course welcome to experiment with it and to draw their own conclusions.

CHAPTER 20
OTHER HARMONIC NUMBERS

So far in this book, we have concentrated entirely on the fourth-, fifth-, seventh- and ninth-harmonic charts. The choice of these particular harmonic numbers is, of course, a matter of personal judgement. In Chapter 2, I argued that the practising astrologer does not have time to work out a large number of harmonic charts for every birth chart that he deals with, and so must restrict himself to a small number of harmonic charts which are likely to be useful for interpretation; and I gave my reasons for believing that the H4, H5, H7 and H9 charts are likely to be the most useful for this purpose.

However, there may well be readers who want to experiment with other harmonic numbers, either for research purposes or because they want to obtain a still deeper insight into a particular birth chart. In this chapter, therefore, I will briefly discuss some other harmonic numbers. The discussion will be tentative, since I do not have extensive experience of these other harmonic charts, and I will mostly be talking at a theoretical level, discussing what these charts *ought* to mean (in the light of what we already know about the meaning of particular numbers), rather than what I have *found* them to mean through empirical investigation. I will also, however, be relying to some extent on the findings of other astrologers.

I will divide these 'other harmonic numbers' into four groups: Two and multiples of Two; Three and multiples of Three; multiples of Five and Seven; other prime numbers beyond Seven.

Two and multiples of Two

The numbers 2, 4, 8, 16 . . . are clearly a series (2, 2^2, 2^3, and so on). Each of them represents the principle of pure Twoness at a particular level. Elsewhere in this book the fourth-harmonic (H4) chart was selected to represent the principle of pure

Twoness; but it may also be worthwhile to consider other harmonic charts in the series.

The **second-harmonic (H2) chart** does not show any aspects that were not in the radical or H4 charts, but it presents the aspects in a different way. In the H2 chart, the oppositions in the radical chart become conjunctions, the squares become oppositions, and the semi-squares become squares. The H2 chart is concerned with the principle of Twoness at a simple and basic level, and this basic Twoness is, I believe, concerned with *the individual's sense of identity in relation to other,* his awareness of himself (or of his self) as a separate entity operating within an environment which is not part of himself. Hence the H2 chart (even though it does not provide any hard information which is not in the other harmonic charts) can be expected to be particularly revealing of the individual's sense of personal identity, his relationship with his environment, and perhaps also his ability to form one-to-one relationships with other people.

When we move on to the **eighth-harmonic (H8) chart**, we have moved beyond the H4 chart and into the third level of Twoness (2^3). The H4 chart is, as we have said, one of *effort*: it shows the efforts that a person makes to cope with his environment. The H8 chart goes beyond this, and can be expected to show the *results of effort,* and hence is a chart of *achievement*. In the H8 chart, the radical semi-squares become conjunctions. These semi-squares, as I have said, are particularly concerned with *manifestation*: so if a person has two planets in a radical semi-square, he will be keen to manifest (or demonstrate) to the world that he is able to bring these planets together and to overcome his difficulties in so doing. We can, therefore, study the H8 chart for indications of the ways in which the individual strives to overcome his problems and to cope with his environment though visible (manifest) action. (However, the H8 chart contains little information that is not in the H4 chart. We can judge the principle of manifestation by studying the oppositions and other 'hard' aspects in the H4 chart.)

The **sixteenth-harmonic (H16) chart** goes beyond this again. If H8 is concerned with manifestation, we would expect H16 to be concerned with the *results of manifestation*. Charles Harvey[1] says that Hindu astrologers regard the sixteenth harmonic as revealing the 'maximum concrete manifestation' of the individual, or 'the fullest manifestation of the individual in the material world'. However we should note that this H16 chart is the one in which the squares in the H4 chart become conjunctions, and that these squares in H4 often seem to represent open conflict, in which one side of the personality is projected on to other people and is regarded as the 'enemy' against whom one must fight. The result of manifestation can often be conflict, since, by manifesting himself to the world in a particular way, a person may come into conflict with another person whose way of manifesting himself is different and incompatible. (I am writing this in the middle of the Falklands crisis of 1982, and thinking of the conflict that has arisen from both Britain and Argentina sticking to their principles. Each side has coped with its internal dilemmas by adopting and manifesting a 'fighting stance' against the outside world, and by identifying an external enemy against whom it must fight in order to maintain its own separateness. This is how the principle of pure Twoness operates.) Hence I think that the H16 chart can also be seen as a chart of conflict. It can show the type of conflict with which the individual is likely to become involved as a result of his efforts to resolve his internal dilemmas and to cope with the challenges from outside.

We could, of course, continue the sequence further into H32, H64, and so on. But my belief is that with H16 we have reached the limit of what is valuable for purposes of interpretation. By drawing aspects on an H16 chart, we obtain an insight into harmonics that are still further along the sequence of Twoness: thus, semi-squares in the H16 chart represent conjunctions in H128, which is 2^7. A conjunction in the 128th harmonic represents an aspect in the radical chart of only 1°52′. We can go beyond this and consider still smaller aspects (2^8, 2^9 and so on), but it is doubtful whether it is worthwhile to do so. Unless we believe that the sequence goes on for ever and that *all* the powers of 2 are equally significant (for instance, that 2^{100} and 2^{1000} are just as significant as 2^2), we have to believe that there is a gradual reduction in significance as the harmonic number increases, so that a high-numbered harmonic chart will have less value for interpretation than a low-numbered harmonic chart.

Three and multiples of Three

The **third-harmonic (H3) chart** is concerned with the principle of Threeness at a simple and basic level. This basic Threeness is, I believe, concerned with the individual's pleasure in realizing that he is part of the world around him. Whereas Twoness is the level at which the individual realizes his selfhood and therefore his separateness (self versus other), Threeness is the level at which he realizes that he is not after all entirely separate, but that he has connections with the outside world. The H3 chart, therefore, can be expected to show the ways in which he explores these connections and *makes contact* with the world about him. Essentially it is a chart of *pleasure,* since the Threeness principle is to do with pleasure and enjoyment in the realization that one fits in with, or harmonizes with, one's environment. Although I do not use the H3 chart, I believe that it could be a most valuable chart for interpretation, and may indeed have advantages over the H9 chart, which shows pleasure at a more refined and less easily attainable level. However, the H3 chart does not contain any aspects which are not in the radical chart or the H9 chart: thus, the conjunctions in the H3 chart are visible as trines in the radical chart, and the trines in the H3 chart are visible as conjunctions in the H9 chart.

If we follow the sequence 3, 3^2, 3^3 . . ., we come first to the H9 chart (which we have discussed elsewhere) and beyond this to the H27 chart, which represents a very pure type of pleasure (bliss or esctasy). On the basis of experience with the H9 chart (in which conjunctions in H27 appear as trines), I do not believe that the H27 chart would have great interpretative value, except perhaps for those rare people who have reached a high level of enlightenment.

However, we can also multiply Three by Two and multiples of Two. This leads us to the sequence 6, 12, 24 . . . . These can be called 'mixed' numbers since they are made up of more than one prime number. They are concerned with striving towards pleasure, or pleasure in striving.

Thus, the **sixth-harmonic (H6) chart** and the **twelfth-harmonic (H12) chart** are concerned with this combination of Threeness and Twoness. Their importance derives from the fact that since Threeness is essentially to do with passive enjoyment, it probably needs an injection of Twoness in order to make it manifest in the world. Thus, John Addey and Charles Harvey[2] say that the H6 chart is to do with 'the objective expression of the number three: the rhythm of life . . . the outward expression of the joy of life . . . the rhythms of work and play . . . connected with all rhythmic activity'. The H12 chart (which contains a further injection of Twoness, and so is concerned with *effort* towards pleasure and connectedness) is 'probably a very important harmonic chart showing the nature of the terrestrial life as a whole in its objective expression'.

However,I confess that, having experimented with these charts (and especially with the H12 chart), I find it difficult to interpret them because of their 'mixed' nature. I find it easier to interpret a harmonic chart in which the conjunction represents a pure (unmixed) number, so that I can look at other aspects in the chart to see how that number is affected by other numbers. For example, the conjunctions in the H12 chart can be seen *either* as trines in the H4 chart *or* as squares in the H3 chart. In the former case, we are looking at a chart which is essentially to do with effort, so that the trines in the chart show how the person obtains pleasure from this effort. In the latter case the chart is essentially to do with pleasure, so that the squares show how the person will strive towards this pleasure. It is less easy to interpret a chart in which the principles of 'effort' and 'pleasure' are inextricably intermingled.

Despite this, I am willing to admit that the H6 and H12 charts probably *are* important (especially in view of the traditional importance of the numbers 6 and 12 in astrology), and that astrologers who wish to experiment with these charts should be encouraged to do so.

Multiples of Five and Seven

Harmonic charts can be drawn up for combinations of Five and Seven with other numbers. For example, the tenth-harmonic (H10) chart represents the combination of Fiveness with Twoness, and the fifteenth-harmonic (H15) chart represents the combination of Fiveness and Threeness. However, these again are 'mixed' numbers, and once again I am inclined to think that the pure or unmixed harmonic chart has more value for interpretation. Conjunctions in H10 are best seen as oppositions in the H5 chart, and conjunctions in H15 are best seen as trines in the H5 chart. Similarly, it seems better to note the oppositions and trines in the H7 chart, rather than to draw separate charts for H14 and H21.

A stronger case can perhaps be made out for the **twenty-fifth-harmonic (H25) chart** which represents 5^2, and so has the same relation to the H5 chart as the H9 chart has to the H3 chart: it represents a purer or more refined level of Fiveness. John Addey found that the twenty-fifth harmonic was prominent in his study of doctors,[3] and it seems reasonable to suppose that strength in H25 (5 × 5 or 'order in order') is connected with the ability to introduce 'orderliness into the order' of life, so that one can, in an orderly and structured way, arrange and rearrange the types of style and stucture that one uses, so as to make them fit the occasion. Less positively, difficulties in H25 could be related to deep-seated obsessive behaviour (the inability to step outside restrictive patterns and structures) and hence to certain physical or mental dis-orders, and this could help to explain John Addey's finding[4] of a connection between H25 and paralytic poliomyelitis. Hence H25 could be a very important harmonic. Even if we do not draw a separate H25 chart, it may be worthwhile to look for quintiles (and perhaps also semi-quintiles) in the H5 chart, to see whether there is strength in H25.

A similar case could be made out for the **forty-ninth-harmonic (H49) chart,** which represents 7^2. However, forty-nine is a far higher number than any of the others we have considered, and the higher the harmonic number, the more inaccurate the positions of the planets (and still more of the Angles) in the chart become. It would seem inadvisable to draw an H49 chart unless we are certain of the accuracy of the birth time; and the planetary positions should if possible be calculated by computer, not by hand.

Other prime numbers beyond Seven

Traditionally, astrologers were concerned with only three prime numbers – One, Two and Three – and with multiples of these numbers. In this book we have introduced two more prime numbers – Five and Seven. But it is possible to go further and look at other prime numbers beyond Seven. The next five prime numbers are 11, 13, 17, 19 and 23. Since each of these is a prime number (not divisible by any number except itself and One), each can be expected to introduce a new principle or quality: thus there must be a quality of Elevenness and a quality of Thirteenness, separate and distinct from the Twoness, Threeness, Fiveness and Sevenness which we have already considered. Is it possible to predict the nature of each of these new qualities?

I believe that, in very general terms, it is possible to do so, by considering the *sequence* of the prime numbers 2, 3, 5, 7 . . ., and considering the *direction* in which this sequence appears to be moving. My suggestion is that it is moving in the direction of *internality*.

The principle of Twoness is concerned with the individual's relationship with the external world. Within the principle of

Twoness, the individual is in fact *dominated* by the external world: he sees it as setting him problems, difficulties and challenges which he must fight to overcome. At the most primitive level, he struggles for survival; beyond this, he struggles for success, but it is success *on the world's terms,* not on his own. His definition of self is dictated by the demands of other: hence the need always to identify enemies against whom one must fight, since without an enemy (or without an external challenge) one has no true identity.

The principle of Threeness represents the ideal balance between externality and internality. The individual feels that he is part of the external world, but he is not dominated by it, so his own individuality is not diminished by it. Also he does not try to dominate it and to impose on it his own definition of reality. Hence Threeness (together with Nineness, which is its higher power) represents true understanding of one's place in the world.

Within the principle of Fiveness, the individual starts to dominate the world and to *rearrange* it to suit his own needs as he himself has defined them. He tries to create an *artificial* world of his own making. He sees himself as the centre of the world, and he sees the world as an extension of himself, ready to be shaped to his needs. Thus, the sources of Fiveness are within the individual himself (instead of being imposed on him from outside, or created out of his sense of belonging in the world).

With the principle of Sevenness, we move still further towards internality. Sevenness is to do with inspiration, but it is inspiration *proceeding from within the individual* rather than received from outside. Within the principle of Sevenness, the individual interprets the world in highly personal ways which seem to have their source in dreams and fantasies within his conscious or subconscious mind. Unrestrained Sevenness often seems to lead to a lack of contact with reality.

Thus we can see that as we move along this sequence we move further and further towards *internality* – towards drives that are initiated within the individual rather than being imposed on him from outside. Hence we would expect that, if we move still further into Elevenness and Thirteenness, we will move still further towards drives that are initiated still deeper within the subconscious mind, and bear less and less relation to the objective nature of a person's relationship with the outside world. Of course, this does not tell us the precise nature of Elevenness and Thirteenness as separate qualities; but it gives us some guidance on the area in which we should be looking.

We can also note that the people who exhibit the qualities of Fiveness and Sevenness most strongly are usually those who are relatively weak in the qualities of Twoness and Threeness. Therefore we can expect that the people who will be strongest in Elevenness and Thirteenness will be those who are relatively weak in *all* of the previous prime numbers.

What then is the precise meaning of Elevenness and Thirteenness? We will take first the **eleventh-harmonic (H11) chart**. Seymour-Smith,[5] who uses H11 aspects regularly in interpretation, says that they indicate 'excess'; he also quotes Williamsen as saying that they describe 'a person's ability to integrate diversities and dualities'. He goes on to say that these two interpretations are reconcilable: 'the tension of "double-bind" situations, which can be external – where you are trapped by feelings of obligation or duty, but cannot entirely please one or more people – or internal – the tug between scepticism and faith – is likely to lead to a type of stress which in turn will lead to excess in one form or another'.

On the basis of my own investigations of H11 – though they are less extensive than my investigations of the earlier harmonics – I would like to suggest a slightly different interpretation. It seems to me that when planets come together in H11, the person has a *deep internal longing* to bring them together: they form part of his *deep-seated internal fantasy* about himself. We have used the word *fantasy* before when dealing with H7, but whereas the fantasies arising from H7 have a wild, fitful, impulsive quality, those arising from H11 seem to have a much more steadily-burning, obsessive, perhaps dream-like quality. They are the stuff of the person's dreams about himself.

This interpretation is not so far removed from Seymour-Smith's as may at first appear. For, if the person's dream-image of himself is strong enough, he may try to translate it into reality through his behaviour; but, since the dream arises from within himself and has no relation to external reality, it is difficult for him to act it out in a realistic way and to pitch it at the right level. Either he will underplay it (in which case it may not be noticed at all), or else he will overplay it and carry it to excess: and there do in fact seem to be many cases where the acting-out of the inner dream is carried to excess. And yet the excess may lead to revulsion, when the person comes to realize that his behaviour has been only the manifestation of an inner fantasy, and resolves to push it down again within himself.

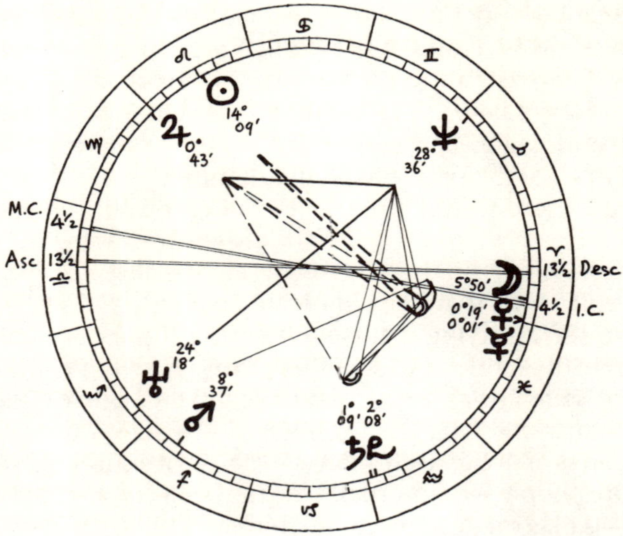

Paul Gauguin: H11 chart

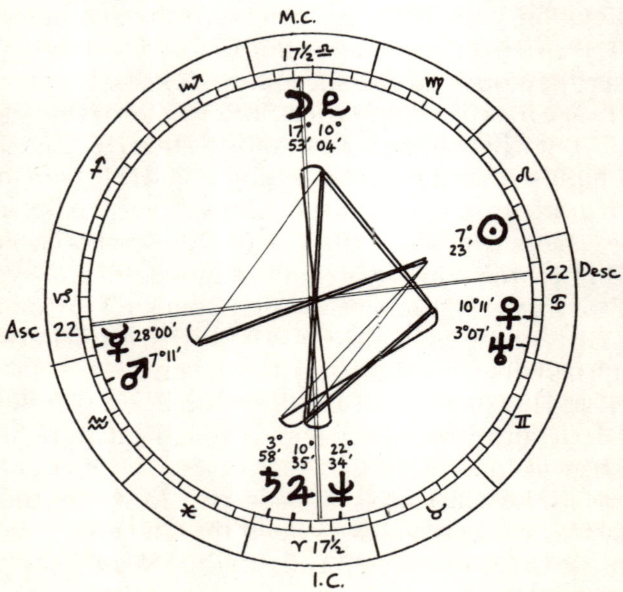

Greta Garbo: H11 chart

Two of the strongest H11 charts that I have seen among famous people are those of the painter **Paul Gauguin** and the actress **Greta Garbo.** Gauguin is lacking in strong configurations in the previous harmonics (H4, H5, H7, H9) so that he is particularly dependent on his H11 pattern of planets; and in his H11 chart (p.262) there is a Mercury-Venus conjunction which is exceptionally strong: firstly from being almost exact, secondly from being close to the Moon, thirdly from being close to both the Descendant and the I.C., and fourthly from being linked with five other planets (Sun, Jupiter, Saturn, Neptune, Pluto). It would seem that, in fleeing to Tahiti, he was pursuing the dream which is represented by this Mercury-Venus conjunction, and his late paintings were an attempt to portray the dream on canvas. Gauguin's behaviour (as seen by his contemporaries) was certainly unrealistic and excessive, and his life in Tahiti seems to have been a desperate attempt to preserve the beautiful dream against the encroachment of harsh reality.

Garbo's H11 chart (p.262) contains (in addition to many other aspects) a very strong Venus-Jupiter-Pluto T-square and an exact Sun-Mars opposition. This is a complex configuration, and we can see that it contains elements *both* of the desire to go on stage and win the adulation of the world, *and* of the desire for privacy (shutting herself off from the world) for which Garbo later became so famous. But the consistent factor is that Garbo was driven by her dreams – her inner fantasies about herself – more than by the objective reality of her situation. She herself said: 'Even when I was a tiny girl, I preferred being alone . . . . I could give my imagination free rein and live in a world of lovely dreams'.[6]

This is a provisional interpretation of the meaning of the eleventh harmonic, and further investigation may well reveal more about its essential nature.

We can now turn to the **thirteenth-harmonic (H13) chart.** Seymour-Smith,[7] following Williamsen, says that the H13 chart 'gives indications of a person's attitude both to death and to spiritual deadness', and this does indeed seem to be part of the truth about the thirteenth harmonic.

I have not studied many H13 charts, but I would like to report on one very remarkable case: that of **Peter Sellers** (whose H9 chart was presented in Chapter 7). I found Peter Sellers's earlier harmonic charts disappointing for, although they explained a great deal about him, they did not explain the central mystery at the heart of his personality, which was that Sellers seemed to

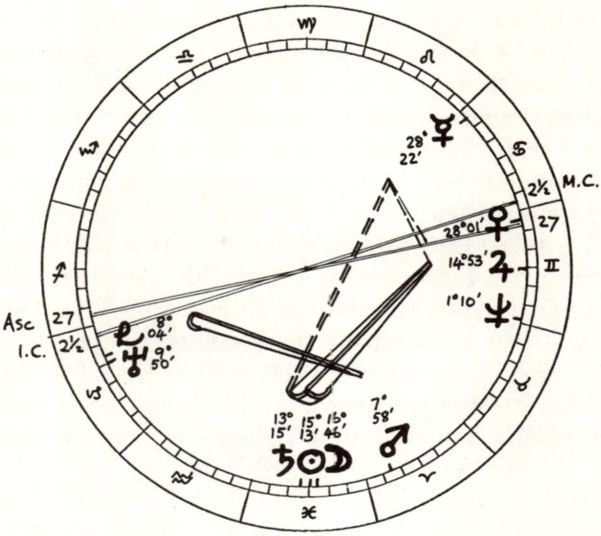

Peter Sellers: H13 chart

have difficulty in believing in his own existence. Robert Parrish, the film director, said of him: 'He walked this strange tightrope of not being a real person at all'; and his friend Peter O'Toole said: 'Pete had an extraordinary sense of not being there. He genuinely felt that when he went into a room no one could see him'.[8] But when I turned to the H13 chart the mystery was solved: for in the H13 chart (above) Sellers has a close conjunction of Sun, Moon and Saturn, all square to Jupiter. If H13 is to do with *one's attitude to one's own existence,* then we can see how this conjunction would cause Sellers (within the H13 mode of thought) to feel so weighed down by Saturn that he felt himself scarcely to have an existence at all. This case, and others that I have looked at, support the view that H13 is concerned with deep-seated attitudes towards one's own existence, and therefore also towards one's own death.

If these interpretations of the eleventh and thirteenth harmonics are correct, we can see how, as we proceed further along the sequence of prime numbers, we are proceeding still further towards *internality*: that is, towards desires, thoughts and feelings which have their origins within the person's mind and are also increasingly introspective (concerned with contemplation of the person himself), so that they bear no relation to the person's objective situation within his environment. We would expect that if we proceed further along the sequence (into H17, H19, H23 and so on), we would be probing still deeper into this internality. However, my own guess would be that, *for most people,* the thoughts and feelings associated with these higher prime numbers are probably buried so deep in the psyche that the person is scarcely aware of them, and still less able to integrate them into his visible behaviour. But there are probably exceptions to this: for instance, a close Sun-Moon-Saturn conjunction (such as Peter Sellers has in H13) would be so powerful that it would be likely to push itself to the surface even if it occurred in H17, H19 or H23. It is probably through the study of these exceptional cases that we will come to understand the meaning of the higher prime numbers.

Conclusion

It is difficult to reach conclusions about which are the most useful harmonic charts for interpretation, and I can only express a personal view. If I were to choose which harmonic charts I most wanted to add to the charts which I already use for interpretation (H4, H5, H7 and H9), my priorities would be (in this order): H11, H3, H13, H16, H12 and H25. In fact, the experience of writing this chapter has made me resolve to use the H11 and H3 charts from now on! The H13 chart also seems promising, but I do not as yet fully understand it. The H16 chart seems useful in uncovering a different level of Twoness, but I am not sure whether it would always reveal information that was not in the H4 chart. I have difficulty with the H12 chart, but I am influenced by the fact that Charles Harvey and others have found it very useful. Finally, the H25 chart could be helpful in some cases, but I would prefer to start by looking for quintiles and semi-quintiles in the H5 chart, and I would draw the H25 chart only if it seemed likely that it would contain a strong configuration.

If we studied all ten of these harmonic charts (in addition to the radical chart), we could obtain a very full understanding of the person's astrological personality; but the amount of work would be considerable, and it is for each astrologer to decide how far he wants to go along this path. Probably most astrologers will want to start by working out these harmonics for their own charts and those of their close friends, so as to see how much insight they do in fact obtain from them.

A computer print-out of the planetary positions in the first thirty or sixty harmonics (such as is obtainable from Astro Computing Services) makes it possible to scan the figures to see which harmonic charts are likely to be especially strong. For instance, a strong conjunction such as Peter Sellers has in H13 would be easily visible in the print-out, and an astrologer could then draw the H13 chart to obtain the full picture, even if he was not in the habit of drawing H13 charts for every case.

But really the greatest need is for investigation and enquiry. In looking at the higher harmonics we are exploring uncharted territory. I hope that many astrologers will venture into this territory and report on what they find there.

CHAPTER 21

HARMONICS OF THE ZODIACAL CIRCLE

The title of this chapter may seem surprising, as we have been dealing with harmonics of the zodiacal circle throughout the book. That is, we have been creating harmonic charts, within which the planets are assigned to their places within the mini-zodiacs which are subdivisions (or harmonics) of the total zodiacal circle.

It is true that we have been doing this: but we have not been attempting to *interpret* the planets according to their position in the harmonic zodiac. Our interpretations have been based on the *aspects* formed by the planets in the harmonic chart, backed up by the zodiacal signs which they occupy in the *radical* chart. If a planet is in Aries in the radical chart, we have treated it as being *really* in Aries in the harmonic charts also, despite the fact that it may *appear* as being (for instance) in Gemini in the H4 chart, in Cancer in the H5 chart, in Leo in the H7 chart, and in Libra in the H9 chart. This chapter, then, is concerned with the question: can we also interpret the signs which the planets occupy in the harmonic charts?

Many astrologers are inclined to answer yes to this question. They believe (for instance) that if a planet falls in Gemini in the H5 chart, it must in some way have the qualities of Gemini, and it must be interpreted as being in Gemini within the H5 chart. For instance, if we regard H5 as being concerned with style, then a person with Sun in Gemini in the H5 chart would have a Geminian style, despite the fact that his radical Sun might be in Aries. And from this it would follow that aspects *between* the harmonic chart and the radical chart would also be important. For instance, if a person has radical Sun at 10° Aries and radical Moon at 10° Taurus, then the Sun in his H4 chart would be at 10° Taurus conjunct his radical Moon. Many systems of interpreting harmonic charts are based on the observation of these aspects (and especially conjunctions) *between* the harmonic chart and the radical chart.

I do not believe that this procedure is justified, since it is based on a theory whose truth has never been demonstrated (though astrologers tend to take it for granted): that the zodiac *really starts* at 0° Aries.

In order to examine this, we must go back to square one and ask ourselves what the zodiac really is.

The zodiac is a division of the ecliptic circle into twelve equal signs (or twelve segments of 30 degrees each). The circle which it divides is of course (like any circle) continuous: it has no starting-point and no ending-point. Yet we can say that there are four points around the circle (at intervals of 90 degrees) which have a special significance: these are 0° Aries and 0° Libra (the equinoctial points) and 0° Cancer and 0° Capricorn (the solstitial points). From a cosmic point of view, it would seem that no one of these points has primacy over the other three. It is understandable that the early astrologers, from their vantage-point in the northern hemisphere, should have chosen 0° Aries (the spring equinox) as the start of the astrological year; but, if they had happened to live in the southern hemisphere, they would have been more likely to choose 0° Libra. A case can also be made out for 0° Capricorn, which is close to the Galactic Centre (26½° Sagittarius), or for 0° Cancer, which is almost *opposite* to the Galactic Centre; and in general there is a strong case for starting the zodiac at one of the solstitial points (where the Sun starts its journey south or north) rather than at one of the equinoctial points (where the Sun does not change direction). Thus, any one of these four points could equally well be taken as the starting-point for the zodiac.

But really there is no reason why the zodiac should have a starting-point at all, since it is a continuous and never-ending cycle of development. Aries follows on from Pisces, in just the same way as Taurus follows on from Aries. For purposes of astrological calculation it is convenient to treat the zodiac *as if* it had a starting-point (just as it is convenient to treat the cycle of terrestrial longitude *as if* it started at the Greenwich meridian), and 0° Aries will do as well as any other point for this purpose; but this is not to say that 0° Aries is the *true* starting-point.

But the signs which the planets occupy in the harmonic charts are dependent on the starting-point which we have chosen for the zodiac. For example, we can take the case that we have already mentioned, in which radical Sun is at 10° Aries and radical Moon is at 10° Taurus. If the zodiac starts at 0° Aries, then the H4 Sun is at 10° Taurus, conjunct the radical Moon. But if the zodiac starts at 0° Libra, then the H4 Sun is at 10° Scorpio, opposite the radical Moon; and if the zodiac starts at 0° Cancer or 0° Capricorn, then the H4 Sun is at 10° Leo or 10° Aquarius, square to the radical Moon. If we believe that the harmonic signs, and the aspects between harmonic planets and radical planets, are significant, then we will interpret the H4 chart differently in each case.

If, however, we accept that the zodiac has no true starting-point, then we are still free to use 0° Aries as a convenient starting-point for calculation (comparable to the Greenwich meridian), but we will not fall into the trap of thinking that Gemini in the H4 chart *means* the same as Gemini in the radical chart, or that a planet at 20° Gemini in the H4 chart is, in any meaningful sense, in conjunction with a planet at 20° Gemini in the radical chart. In drawing up the H4 chart, we are creating a mini-zodiac whose total length is 90 degrees: we are at liberty to divide this zodiac into twelve equal segments and to give these segments the labels Aries, Taurus, Gemini and so on, but *we do not know what these segments mean until we have investigated them within this particular harmonic.* Each of the harmonic zodiacs must refer to some type of cycle, related to the quality of that particular harmonic; but the nature of the cycle must be investigated separately in each case. It does not necesarily follow an Aries-Taurus-Gemini sequence as these signs are generally understood.

In fact, in the case of the H4 chart, we do know something about the nature of the cycle. All of the cardinal signs in the radical chart are converted into the first four signs (Aries, Taurus, Gemini, Cancer) in the H4 chart; all of the fixed signs are converted into the next four signs (Leo through Scorpio) in the H4 chart; and all of the mutable signs are converted into the last four signs (Sagittarius through Pisces) in the H4 chart. Therefore the cycle of the H4 zodiac is the cycle from cardinality through fixity and mutability and back to cardinality. Moreover, Aries in the H4 chart refers to the first 7½ degrees of the radical cardinal signs; Taurus in the H4 chart refers to the next 7½ degrees, and so on. Thus, in investigating whether H4 Aries is different from H4 Taurus, we are in fact investigating whether the meaning of cardinality varies according to the degree which the planet occupies in the radical chart.

Similarly, the cycle of the H3 zodiac is the cycle from Fire through Earth, Air and Water and back to Fire; the cycle of the H6 zodiac is the cycle from positivity to negativity and back to positivity; and the cycle of the H12 zodiac is the cycle from the beginning to the end of a sign in the radical zodiac.

But in the case of other harmonics – including H5, H7 and H9 – traditional astrology tells us nothing about the meaning of the zodiacal cycle, and we can only discover its meaning by empirical research. For instance, since H7 is connected with inspiration, we would expect that the quality of inspiration of a planet in H7 Aries would be different from that of a planet in H7 Virgo; but we cannot necessarily assume that the former has an Aries quality and the latter a Virgo quality.

In fact, the research which would establish the true nature of these harmonic cycles has not yet been fully carried out. But the research which *has* been done seems to show that the true picture is extremely complex, and that it would be optimistic to expect that we will ever be able to identify a single characteristic which corresponds (for instance) to Virgo in the H7 chart. Several researchers have taken the natal Sun-positions of a large number of people within a profession (doctors, artists, clergymen, etc.), and have subjected these Sun positions to harmonic analysis. One of the best-known of these is John Addey's study[1] of 7,302 doctors of medicine (based on previous

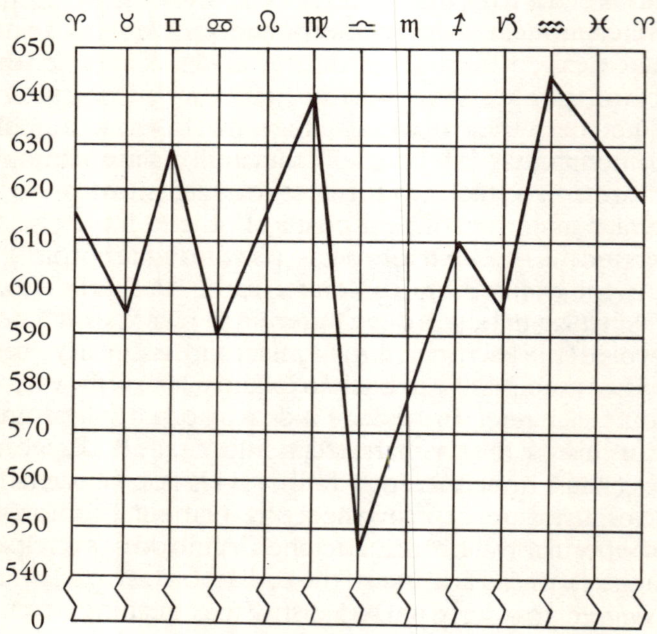

Fig. 10 Distribution of Suns of 7,302 doctors in the H5 Zodiac

work by R. C. Firebrace and Rupert Gleadow). Figure 10 (p.270) shows the Suns of these 7,302 doctors distributed among the twelve signs of the fifth-harmonic zodiac[2]. It seems to show that doctors of medicine tend to have Sun in Aquarius or Virgo in the H5 chart, and tend *not* to have Sun in H5 Libra or Scorpio. But John Addey shows that the situation is in fact more complex than this. The distribution of the doctors' Suns through the zodiac shows a strong wave-pattern not only in H5 but also in several subdivisions of H5 such as H10 and H25. Similarly, John Addey's study of the Sun-positions of 4,465 clergymen[3] (based on the combined samples of Donald Bradley in the USA and R.C. Firebrace in the UK) shows that there were strong wave-patterns not only in H7 but also in several subdivisions of H7 such as H21 and H49.

Another example (though not based on a statistical survey) is my own observation[4] that the great romantic composers (Beethoven, Berlioz, Debussy, Ravel, Schumann) seem to tend to have the outer planets (especially Uranus and Neptune) in the second half of the cardinal signs in the H7 chart.

The problem with such findings, of course, is what do they mean, and how can we use them in interpretation? At present the answer has to be that we do not know what they mean, and so we cannot use them in interpretation. We would not (for instance) expect there to be a direct causal link between the H5 chart and the question of whether its owner becomes a doctor of medicine: rather we would expect that doctors of medicine would tend to have a number of characteristics in common with one another, and that some of these characteristics may possibly correlate with the zodiacal position of the Sun (or of other planets) in the H5 chart. In order to discover what Sun in Gemini (or in a particular degree of Gemini) in the H5 chart *means,* we will have to study these underlying characteristics. The keywords approach, which Gauquelin and Addey have applied to the harmonics of the diurnal circle (and which we will describe in the next chapter), would seem also to be the most promising approach to the study of the harmonics of the zodiacal circle; and we must hope that such a study will soon be carried out. Yet we must also accept that the secrets of the harmonic zodiacs may be peculiarly difficult to decipher. John Addey's work seems to suggest that each degree of the zodiac, or even each fraction of a degree, may have its own distinctive meaning because of its position on a very large number of harmonic cycles; and the unravelling of these separate meanings through empirical research will be extremely difficult and laborious. Computers, of course, are able to do an almost infinite amount of work; but we may not yet be able to feed the computers with enough data or to ask them the right questions.

At present, therefore, an astrologer who is doing interpretative work must ignore the signs in which the planets and Angles are placed in the harmonic charts, since we do not know what these signs mean. The harmonic charts should be viewed as patterns of aspects, showing the nature of the Fourness, Fiveness, Sevenness or Nineness in the person's make-up. But the planets in the harmonic charts are still the same planets that were in the radical chart, and they should be treated as being in the same sign (and also the same house) that they occupied in the radical chart.

In interpreting the signs in the radical chart, we must at present rely on traditional astrology. But eventually, harmonics may show us the way to clearer and fuller understanding of the zodiacal signs. Almost certainly, they will show that the signs are not separate and discrete 'boxes' (as astrologers have traditionally viewed them), but that they merge gradually into one another. John Addey says: 'The view which has been forced upon me by a long study of statistics of zodiacal positions, mundane positions, aspects and other astrological factors is that we are dealing, not with boxes, but with *waves*'.[5] Traditionally, astrologers have interpreted 29° Aries as being identical with 0° Aries, but as quite different from 0° Taurus. But this is quite inconsistent with harmonic theory, which says that there are no sharp dividing-lines but only waves which gradually rise and fall. If harmonics have any validity, it *must* be the case that 29° Aries has more in common with 0° Taurus than it has with 0° Aries. As yet we do not know precisely how to interpret 29° Aries; but we can perhaps use our common sense and realize that, if the traditional meanings of the signs have any validity, then both 29° Aries and 0° Taurus must be transitional stages between 'pure' Aries and 'pure' Taurus. In other words, the qualities associated with these degrees must be a blending of the qualities of Aries with the qualities of Taurus.

At the Astrological Association's conference in Nottingham in 1976, I presented a paper (which has never been published) on 'Overlapping signs and houses'. The theme of this paper (so far as the signs are concerned) was that as we proceed through each sign from 0° to 30°, the qualities of the sign (which start as external manifestations) become gradually more internalized, so that by the end of the sign they take the form of internal attitudes rather than external behaviours. The external behaviours associated with the *next* sign then begin to arise out of the internal attitudes associated with the *previous* sign.

Thus, for example, the last degrees of Pisces and the first degrees of Aries are both associated with a type of aggressive and confrontational behaviour (Aries manifesting on the surface) which seems to arise out of an inner Piscean attitude of sensitivity and vulnerability. However, as we proceed further through Aries, the sensitivity and vulnerability become less apparent and the confidence and assertiveness of Aries become more dominant. By the time we reach the second half of Aries, the qualities of Aries have started to manifest themselves not as external aggressiveness but as internal confidence and single-mindedness. Before we reach the end of Aries, this inner confidence has started to be overlaid by external behaviour-patterns which have the qualities of Taurus (stubbornness, obstinacy, sensuousness): this Taurean behaviour is possible only because of the Arien confidence which underlies it.

Based on this theory, I presented a scheme for the interpretation of the different degree-areas within each sign, as follows:

Cusp (around 0°)
I have become dissatisfied with my passive view of the world, and aware of the need for pro-active behaviour; I learn a new method of self-projection (in the manner of the later sign) with which I partly cover up my inner feelings about the world (which are in the manner of the earlier sign).
Early sign (around 7½°)
I concentrate mainly on outward self-projection, imposing myself on the world (in the manner of the sign).
Middle sign (around 15°)
I become dissatisfied with self-projection, and aware of the need for reactive behaviour: I learn a new method of seeing and responding to the world (in the manner of the sign) which makes me feel more at home in my environment.
Late sign (around 22½°)
I concentrate mainly on inward gathering of impressions of the world, and reflecting my perceptions out to the world again (in the manner of the sign).

This scheme is in effect a proposal for the interpretation of the phases of the twelfth harmonic of the zodiacal cycle. Although some empirical evidence was presented in the paper, it remains in fact an untested theory (though I believe that a similar theory has been developed by Bruno and Louise Huber[6]). Readers are welcome to experiment with it as they wish. In the long run, however, such a simplistic theory is likely to be overtaken by the more detailed understanding of signs and degree areas which will emerge from harmonic research.

Planetary harmonics

I should also mention the technique of 'planetary harmonics', which has been invented by John Greig.[7] Greig suggests that each planet in a birth chart 'vibrates' to the harmonic which is associated with its position in the zodiac (starting at 0° Aries). Thus, for example, if the Sun is at 22°15′ Leo (for which the absolute longitude, converted from minutes into decimals, is 142.25°), then the Sun 'vibrates' to the harmonic which is represented by 360 divided by 142.25. Greig then suggests that it is possible to construct a planetary harmonic chart (or P.H. chart) for the Sun, in which the zodiacal position of each of the other planets (and also of the Ascendant and M.C.) is multiplied by this fraction. For instance, if the radical Moon was at 23°45′ Virgo (absolute longitude 173.75°), then the position of the Moon in the Sun P.H. chart would be:

$$\frac{360}{142.25} \times 173.75 - 360 = 79.72 = \text{Gemini } 19°43'$$

Note: 360, or a multiple of 360, must be subtracted if necessary in order to make the result less than 360.)

P.H. charts can be drawn up for each of the planets, and also for the Ascendant and M.C., using this procedure, and Greig claims that each P.H. chart says something about the operation of its own planet. For instance, the Mercury P.H. chart describes mental and communicative faculties. (We should also note that Greig places the 'home planet' – for instance, Mercury in the Mercury P.H. chart – at its original radical position, although according to his own logic the 'home planet' should always be at 0° Aries.)

This is a very odd procedure. John Addey, in his Foreword to Greig's booklet, says that it is 'a perfectly sound procedure, symbolically speaking, for which good precedents already exist in astrological practice'. To me, however, it is unacceptable, for the following reasons:

1 Fractional harmonics do not seem to make sense. Throughout the rest of this book we have been dealing with whole-number harmonics (the fourth harmonic, the fifth harmonic, etc). But the harmonic of 360 divided by 142.25 is the 2.53076th harmonic; and within this harmonic we are creating a mini-zodiac, in the same way as we did in the whole-number harmonics. This mini-zodiac fits into the original zodiac not two times but 2.53076 times. In other words, we have divided the original zodiac into three unequal segments: the first two segments contain the complete mini-zodiac from Aries to Pisces, but the third segment contains the mini-zodiac only as far as 11°07′ Libra, at which point we suddenly revert to 0° Aries. This sudden jump seems completely out of keeping with everything we know about how astrology (and especially harmonic astrology) works.

2 Planetary harmonics are completely dependent on the assumption that the zodiac *really starts* at 0° Aries; and, as we have already shown, this assumption is highly suspect. (It is true that Greig suggests that P.H. positions could be calculated from the Ascendant instead from 0° Aries, but this would be an even more questionable procedure, since it would mean the creation of a mini-zodiac in which the Ascendant was always at 0° Aries, and whose signs were therefore determined by their relationship to the diurnal circle rather than to the original zodiac. This would be a very strange mixing of the zodiacal and diurnal circles.)

3 Even if we accept that 0° Aries is the true starting-point of the zodiac, we may still be unable to accept Greig's account of the strange things that happen at 0° Aries. According to Greig's theory, the harmonic number corresponding to seven successive minutes of the zodiac is as follows:

Pisces 29°57′	1.000139
Pisces 29°58′	1.000093
Pisces 29°59′	1.000046
Aries 0.00′	Infinity
Aries 0°01′	21600.0
Aries 0°02′	10800.0
Aries 0°03′	7200.0

That is to say, at Pisces 29°59′ the harmonic zodiac is almost exactly the same as the radical zodiac, although there is a tiny

> additional segment (1/21600th of a zodiac) tacked onto the end: but at Aries 0°00′ there is suddenly an infinite number of harmonic zodiacs (so that no P.H. chart is possible), and at Aries 0°01′ there are 21,600 harmonic zodiacs (or one for each minute of the radical zodiac). In other words (if we accept Greig's theory of 'vibration') the planet is vibrating 21,599 times as fast at Aries 0°01′ as it is at Pisces 29°59′. Greig's theory seems to suggest that the planet receives a massive input of astrological energy every time it passes 0° Aries. It prodigally expends most of this energy within the first degree of Aries, and has expended nearly all of it before it reaches the end of Aries: thereafter it becomes gradually more and more sluggish as it proceeds round the zodiac, until it once again reaches 0° Aries and its next sudden outburst of energy. (If, however, the planet is proceeding *retrograde* across 0° Aries, the process happens in reverse: an increasingly feverish build-up of energy which then suddenly vanishes.) It is impossible to reconcile this conception either with traditional astrology or with John Addey's harmonic astrology (which is concerned with waves rising and falling, rather than with sudden transformations).

For these reasons I am unable to accept the theory of planetary harmonics. In the end, however, the test of any astrological theory is 'does it work?', and if research should show that planetary harmonics were unquestionably valid, I would have to revise my view, even though the acceptance of this theory would make it necessary for me to rethink my entire conception of how the cosmos works. New techniques (unless they are entirely consistent with the old techniques) cannot be unquestioningly accepted: we must first try to *understand* them by fitting them into our overall model of how astrology works. This is a question to which I shall return in Chapter 23.

CHAPTER 22
HARMONICS OF THE DIURNAL CIRCLE

Last, but most certainly not least, we turn to the harmonics of the diurnal circle, which is the circle of the mundane houses. I have left this to the last because it is not directly related to the study of harmonic *charts,* which is the main subject of this book. However, the diurnal circle is the area of harmonics in which most research has been done, and the most remarkable findings made, and which promises most completely to revolutionize our understanding of astrology. The main researchers in this field have been Michel Gauquelin and John Addey, and this chapter will consist of a brief summary of their work and a discussion of its implications for practising astrologers.

Nature of the diurnal circle

First of all, it must be clearly understood that the diurnal circle is a quite different type of circle (or cycle) from the zodiacal (or ecliptic) circle with which we have been concerned in the rest of this book. Essentially the diurnal circle is a circle, not of space, but of time.[1] Its nature can best be understood by considering the passage of the Sun through the heavens. Each day at a particular time the Sun rises above the horizon; at a later time, it sets. Half-way between these two times, it reaches its upper culmination (its highest point in the sky). This point (the M.C.) and its opposite point (the I.C.) are treated as being square to the horizon (the Ascendant-Descendant axis) in the diurnal circle, *even though they may not be square to the horizon in the circle of the ecliptic.*

In the Placidus house system, the diurnal circle is divided into sectors by dividing the *time taken by each degree of the ecliptic* to pass from lower culmination to the horizon and from the horizon to upper culmination. The system used by Gauquelin and Addey is slightly different. Since they are concerned only with the positions of the *planets,* they have divided the *time taken by the planets themselves* to pass between the Angles. This system (which we can call the Gauquelin system of house division, even though Gauquelin does not believe in twelve discrete houses) gives results which are similar to those of Placidus, but not identical. The differences between Placidus and Gauquelin arise from the fact that, for most of the time, the planets (apart from the Sun) do not lie exactly on the ecliptic, but lie some distance on either side of it. (This is the problem of celestial latitude, which we will discuss in Chapter 23.) Unfortunately it is not possible from an ordinary ephemeris to calculate the positions of the planets according to the Gauquelin system of division (though it is now possible to obtain these positions by sending the chart to Astro Computing Services[2]). But the differences between Placidus and Gauquelin are not very great, and it is possible from a Placidus chart to obtain a broadly correct picture of the positions of the planets within Gauquelin's sectors.

Gauquelin's research

Gauquelin's work (together with that of his wife Françoise Gauquelin) is already very well known among astrologers, and has been very fully reported on by the Gauquelins themselves[3] and by others.[4] Hence I will only provide the briefest of summaries here.

Gauquelin studied the diurnal positions of the planets in the charts of a large number of eminent professional people. The names of the people were obtained from biographical works of reference, and their birth times were obtained from birth certificates. He found that people who are eminent in a particular profession tend (more often than one would expect in a random distribution) to have certain planets in the ninth and twelfth houses, and also (to a lesser extent) in the third and sixth houses. In other words, he found a fourth-harmonic distribution, with four peaks and four troughs around the diurnal circle. Writers tended to have Moon, Venus and Jupiter in these houses; politicians tended to have the Moon; athletes had Mars; actors had Jupiter; soldiers had both Mars and Jupiter; and doctors and scientists tended to have Saturn in these houses.

On the other hand, Gauquelin also found that people eminent in certain professions tended *not* to have certain planets in the third, sixth, ninth and twelfth houses. Thus, Moon in these houses was infrequent (or less frequent than one would expect in a random distribution) for athletes and soldiers; Mars was infrequent for artists and writers; Jupiter for doctors and scientists; and Saturn for actors and writers.

Gauquelin's later work has been concerned with the study of *character-traits*.[5] From biographies and biographical sketches he has extracted all the words which were descriptive of the person's personality, and has then compiled lists of people who had particular character-traits applied to them and has compared this with the diurnal distribution of planets in their birth charts. From this he has found that people having certain character-traits tend again to have certain planets in the third, sixth, ninth and twelfth houses. Gauquelin in his book *Spheres of Destiny* gives lists of the character-traits which correspond to particular planets in these houses: these lists are very long, and we will give here only the character-traits which start with the letter A, since these on their own give a good picture of the type of personality which is being described.

Moon in third , sixth, ninth and twelfth houses: Accessible, accommodating, affable, affected, agile, attentive.
Venus in the same houses: Affable, amiable, attractive.
Mars in the same houses: Active, adventurous, aggressive, agitated, alert, angry, ardent, audacious.
Jupiter in the same, houses: Adventurer, amusing, angry, argumentative, assured, at ease, audacious, authoritative.
Saturn in the same houses: Anxious, assiduous, attentive, austere.

Gauquelin now claims that the links between planets and professions, which were shown in his earlier work, are a by-product of the links between planets and personality. The reason why scientists tend to have Saturn in the ninth or twelfth houses is that they tend to have a 'Saturn-personality' (anxious, assiduous, etc). But if one finds a particular scientist who has more of a 'Jupiter-personality' (adventurous, amusing, etc), then he is more likely to have Jupiter than Saturn in the ninth or twelfth house.

For practising astrologers, the main value of Gauquelin's findings is that they prove, beyond any shadow of doubt, that there is some truth in astrology. Until Gauquelin did his research, there was not a single astrological statement that had been scientifically verified, and it could always be argued that astrologers (and their clients) were simply the victims of their own credulity and gullibility.[6] But Gauquelin has produced results whose statistical significance is so strong, and which have been replicated in so many studies, that there is no possibility that they were produced by chance. Determined teams of scientists have done their best to disprove Gauquelin's conclusions, but have failed. Thus there is now no doubt that there is *some* link between a person's personality and the planets in his birth chart.

But beyond this, Gauquelin's results provide little help to the astrologer who is trying to make a detailed interpretation of a birth chart. Indeed, their main value is a negative one: they show some things that are *not* true, but they provide little guidance on what to put in their place. Thus, for example, Gauquelin proves that people with Mars in the twelfth house tend to be 'active, adventurous, aggressive, angry, audacious', and tend to become soldiers and athletes, whereas people with Mars in the first house tend *not* to have these qualities and *not* to become soldiers and athletes. This is the reverse of the traditional astrological teaching, as given for example by Sakoian and Acker:[7]

> Mars in the First House indicates aggressive, outgoing people who have abundant energy. The physical body is often robust and muscular, giving an appearance of strength and ruggedness. They are not content to be bystanders in life; they must be directly involved in action . . . . This position of Mars gives a love of sports and other forms of physical action.

whereas:

> Mars in the Twelfth House indicates desires and actions that are strongly influenced by the unconscious mind. Work and other activities will be carried on in secret or seclusion. There can be a tendency to act in secret to avoid the open opposition of others . . . . Often they work in large institutions where they can disguise or lose their personal identities.

Quite clearly, we must reject Sakoian and Acker's interpretation, because it cannot be reconciled with what Gauquelin has *proved* to be the case. But what alternative interpretations can we devise for Mars in the first and twelfth houses? Gauquelin does not tell us. All he does is divide the houses into 'areas of high and low intensity': thus, the twelfth house is identical with the third, sixth and ninth houses as being an 'area of high intensity' (except that the intensity is even higher in the ninth and twelfth houses than in the third and sixth) and the first house is identical with houses 2, 4, 5, 7, 8, 10 and 11 as being an 'area of low intensity'.

In various other ways, too, Gauquelin has produced negative results which astrologers can ill afford to ignore. For instance, he has consistently shown that there is no straightforward link between personality (as measured by the character-trait method) and the house positions of Sun, Mercury, Uranus, Neptune and Pluto. More recently, in a study of murderers and of people suffering from various mental illnesses (schizophrenics, manic-depressives, alcoholics, etc),[8] he has shown that there is no clear link between these conditions and the house positions of any of the planets.

Also, the boundaries between Gauquelin's 'sectors' do not correspond exactly to the traditional house cusps. Thus, for example, Gauquelin's equivalent of the 'twelfth house' (although he does not call it this) starts slightly *below* the twelfth cusp and ends slightly *below* the Ascendant. Thus, Gauquelin's work throws doubt on the traditional doctrine of twelve houses starting from the Ascendant, with the Ascendant as the first-house cusp. (But Gauquelin's findings give far more support to the Quadrant house systems, and especially to Placidus, than they do to the Equal house system, which is impossible to reconcile with Gauquelin's findings.)

Thus, Gauquelin has simultaneously proved that astrology is (to some extent) true, and has undermined some of its foundations. As far as the diurnal circle is concerned, we have no alternative but to try to build up a new system of interpretation which is based on empirical evidence and is consistent with Gauquelin's findings. This is what John Addey, before his untimely death, was trying to do .

John Addey's research

Addey, like Gauquelin, used the character-trait method; but, unlike Gauquelin, he compiled lists of character-traits that were prominent at *every* phase of each harmonic cycle in the diurnal circle. (The word 'phase' refers, in effect, to the area surrounding a particular point on the circumference of a circle. In the first harmonic, we can regard the Ascendant as 0°, the I.C. as 90°, the Descendant as 180°, and the M.C. as 270°. But in the second harmonic, we are treating each half-circle as if it was a whole circle: thus the Ascendant and the Descendant are both 0°, the

M.C. and the I.C. are both 180°, and the half-way points between the Angles are either 90° or 270°. Thus, if we say that a particular character-trait is prominent at phase 180° of the second harmonic, we are saying that it is prominent in the area around both the M.C. and I.C., and we are also implying that it is *not* prominent in the areas around the Ascendant and the Descendant.)

Addey has written four articles in the *Astrological Journal* summarizing his findings. In the first two of these articles,[9] he reports mainly on the fourth-harmonic cycles of Saturn, Jupiter and Mars, showing that each of these planets has a different meaning at each phase of the H4 cycle. He supports Gauquelin's finding that the traditional qualities of each of these planets are strongest when the planets are in the third, sixth, ninth and twelfth houses; but he also adds to Gauquelin by describing the qualities that are associated with each of the other phases in the H4 cycle. However, Addey himself says:

> What we have done so far is to isolate the fourth harmonic and to show what this harmonic is about *on its own*. But the full picture cannot be seen until the effects of other harmonics are added. . . . Therefore the practising astrologer should not at this stage attempt to apply these character traits directly to the chart interpretation, or at least only in a broad and general way. We have been looking at one effect among many.[10]

However, in his last two articles,[11] Addey presented detailed findings for the first four harmonics of Mars, and these are presented so fully and clearly that they can be directly used in interpretation, and amount to a new scheme for the interpretation of Mars in the diurnal circle. We will therefore concentrate on Mars for the remainder of this chapter.

(Before going on to summarize Addey's findings, I must explain a difference between Addey's presentation and my own. Addey, following Gauquelin, proceeds round the diurnal circle from the Ascendant in a *clockwise* direction, starting with the twelfth house, this being the direction which the planets actually follow through the sky. I have reverted to the old astrological custom of proceding in an *anti-clockwise* direction, starting with the first house – partly because this will be more easily understood by astrologers, but also because the *transiting* planets in a birth chart move in an anti-clockwise direction, so that this is the direction of movement which is more important in astrology.)

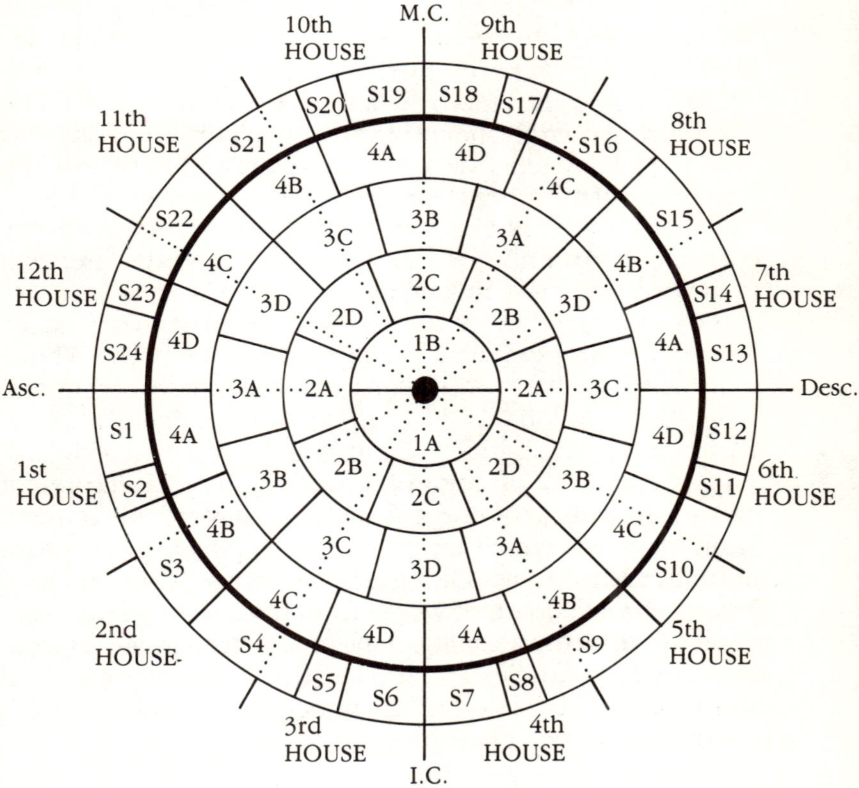

Fig. 11 Phases of Mars in the diurnal circle

Addey discovered significant results for each of the first four harmonics of Mars: that is to say, in each of these harmonics he identified character-traits which are prominent at a particular phase, and which are different from the character-traits which are prominent at the opposite phase. In the case of H2, H3 and H4, he has discovered four significant phases: two pairs of opposing phases, at right angles to each other.

Figure 11 (p.283) shows how these phases are arranged around the diurnal circle. In the innermost circle, 1A and 1B refer to the two opposing phases of the first harmonic (H1). In the next circle, 2A, 2B, 2C and 2D refer to the four distinct phases of H2, and so on. Also shown are the cusps of the astrological houses (using the Gauquelin system which, as we have said, is closer to Placidus than to any of the other commonly-used house systems). Note that the boundaries between the phases (or sectors) rarely correspond to the house cusps: in fact, they *never* do so except in the case of the Ascendant, Descendant, M.C. and I.C.

It should be emphasized that, in every case, the phase-characteristics are strongest near to the *centre* of the phase. This is especially important for H1, since only two phases are shown. Thus, Addey found that the character-traits assigned to phase 1A are strongest in the area within 45° on either side of the I.C.

In describing the meaning of each of the phases, I will present only the six most prominent character-trait words in each case (omitting words which appear in a higher position for a phase of a different harmonic). Only in one case (phase 3C) have I found it necessary to add other words from further down the list in order to clarify the meaning of the phase. Readers are referred to Addey's two articles[11] for a longer list of character-traits and also for a full discussion of the underlying meaning of each harmonic and each phase.

Mars in the diurnal circle: Addey's phases

First harmonic

Phase 1A (90°) (i.e. Mars above the horizon) Charming, popular, witty, comic, gracious, natural.

Phase 1B (270°) (i.e. Mars below the horizon) Powerful, passionate, tenacious, vigorous, rough, critical sense.

Second harmonic

Phase 2A (0°) Common sense, hardworking, has facility, efficacious, straight, true.

Phase 2B (90°) Gay (in the old sense), tender, merry, romantic, good-tempered, amiable.

Phase 2C (180°) Simple, likes family life, lucid, cultivates friendships, tranquil, sincere.

Phase 2D (270°) Reserved, clear, steadfast, independent, cold, proud.

Third harmonic
Phase 3A (0°) Modest, enthusiastic, noble, vitality, honest, humane.

Phase 3B (90°) Good, great-hearted, wise, kind, solitary, fiery.

Phase 3C (180°) Creative, spirited, cultured, imagination, fancy, attends to detail. (The list also includes: popular, charitable, comic, good-tempered.)

Phase 3D (270°) Authority, political, influences others, trustworthy, open, rigorous.

Fourth harmonic
Phase 4A (45°) Precise, calm, reflective, unsociable, likes country life, silent.

Phase 4B (135°) Sensitive, shy, loved, temperate, serious, scrupulous.

Phase 4C (225°) Intelligent, successful, dynamic, optimistic, amusing, brisk.

Phase 4D (315°) Energetic, audacious, frank, hard, active, courageous. (This is the phase which corresponds to Gauquelin's 'area of high intensity'.)

A word needs to be added about the third harmonic, which is plainly different from the others. Addey suggests that 'each phase of the third harmonic . . . may be seen as exemplifying an archetypal life-pattern which must impose itself upon the native as an ideal model'.[12] He suggests the following titles for the archetypes represented by the H3 phases of Mars:

Phase 3A The Hero or Pilgrim
Phase 3B The Knight-Errant
Phase 3C The Truth-Teller or Troubadour
Phase 3D The Governor or Ruler.

I would suggest also that these H3 phases are related to the type of approval which the native seeks from other people through his Mars-behaviour. Thus, Mars in phase 3A says 'I want your admiration'; Mars in phase 3B says 'I want your love'; Mars in phase 3C says 'I want your applause'; and Mars in phase 3D says 'I want your respect and trust'. In some ways 3A is more similar to 3C than to 3B or 3D: this is evidence of a *sixth*-harmonic pattern at work.

If we look again at Figure 11 (p.00), we see that there are twenty-four sectors (marked from S1 to S24) which are distinguished from each other by different combinations of the phases in the four harmonics. We can go through the list of sectors and give a tentative interpretation of Mars in each sector, based on Addey's findings:

Sector 1 (first half of first house) 1A (weak), 2A, 3A, 4A. Hardworking, unsociable, seeking quiet achievement without 'show'.

Sector 2 (third quarter of first house) 1A (weak), 2A, 3B, 4A. Similar to Sector 1 but with greater warmth.

Sector 3 (last quarter of first house, first half of second house) 1A (weak), 2B, 3B, 4B. Kind, good-tempered, sensitive, serious: a very 'caring' position for Mars.

Sector 4 (second half of second house, first quarter of third house) 1A, 2B, 3C, 4C. Romantic, popular, amusing, optimistic, dynamic, liking to please. As Addey says, this position of Mars is the epitome of the 'Great Lover'.

Sector 5 (second quarter of third house) 1A, 2C, 3C, 4D. Similar to Sector 6, but concerned less with power and more with popularity.

Sector 6 (second half of third house) 1A, 2C, 3D, 4D. Energetically devoting oneself to small matters, over which one seeks authority and control: sincere and trustworthy: wanting to be a 'big fish in a small pool'.

Sector 7 (first half of fourth house) 1A, 2C, 3D, 4A. Similar to Sector 6, but with less energy and more reflective calm: quietly managing one's own affairs, wanting family security.

Sector 8 (third quarter of fourth house) 1A, 2C, 3A, 4A. Similar to Sector 7, but concerned less with power and more with personal morality.

Sector 9 (last quarter of fourth house, first half of fifth house) 1A, 2D, 3A, 4B. Proudly, sensitively, charmingly presenting oneself as a person of high moral standards.

Sector 10 (second half of fifth house, first quarter of sixth house) 1A (weak), 2D, 3B, 4C. Kind to others from a position of optimistic and self-confident independence: bestowing bounty from on high.

Sector 11 (second quarter of sixth house) 1A (weak), 2A, 3B, 4D. Similar to Sector 12, but with more warmth and less desire for applause.

Sector 12 (second half of sixth house) 1A (weak), 2A, 3C, 4D. Energetically devoting oneself to hard and creative work in order to win others' approval.

Sector 13 (first half of seventh house) 1B (weak), 2A, 3C, 4A. Similar to Sector 12 but with less energy and more reflective calm: seeking the approval of others for one's solid achievements.

Sector 14 (third quarter of seventh house) 1B (weak), 2A, 3D, 4A. Similar to Sector 13 but with less desire for praise and more for power: seeking authority over others by hard work.

Sector 15 (last quarter of seventh house, first half of eighth house) 1B (weak), 2B, 3D, 4B. Shyly and romantically seeking power: perhaps seeking to compensate for one's deficiencies by obtaining influence over others.

Sector 16 (second half of eighth house, first quarter of ninth house) 1B, 2B, 3A, 4C. A passionate, cheerful, optimistic, high-spirited campaigner for all kinds of causes.

Sector 17 (second quarter of ninth house) 1B, 2C, 3A, 4D. Similar to Sector 18, but with less desire for love and more for admiration.

Sector 18 (second half of ninth house) 1B, 2C, 3B, 4D. Energetically and passionately seeking simple values and the love of one's fellow-men: quick to act in courageous defence of one's values, one's possessions, one's friends and one's way of life.

Sector 19 (first half of tenth house) 1B, 2C, 3B, 4A. Similar to Sector 18 but with less energy and more reflective calm: tenaciously defending what one holds dear.

Sector 20 (third quarter of tenth house) 1B, 2C, 3C, 4A. Similar to Sector 19 but with less desire for love and more for applause.

Sector 21 (last quarter of tenth house, first half of eleventh house) 1B, 2D, 3C, 4B. Tenaciously, but with shyness and reserve, seeking the applause of others for one's creative achievements.

Sector 22 (second half of eleventh house, first quarter of twelfth house) 1B (weak), 2D, 3D, 4C. Optimistically and steadfastly seeking influence over others: reserved and distant, but ingenious and resourceful.

Sector 23 (second quarter of twelfth house) 1B (weak), 2A, 3D, 4D. Similar to Sector 24, but with less desire for admiration and more for power.

Sector 24 (second half of twelfth house) 1B (weak), 2A, 3A, 4D. Energetically and unstintingly seeking people's admiration for one's hard work and one's commonsense efficiency: untiring effort.

These tentative interpretations (which will undoubtedly be improved as astrologers gain more experience with using Addey's findings) are put forward as a basis for interpreting the position of Mars in the diurnal circle. In many cases, however, it will be found that Mars does not lie clearly in any one of the sectors, but lies somewhere on the boundary between them. In such cases we should attempt to reach an interpretation which blends the characteristics of the two adjacent sectors. Addey's and Gauquelin's work shows that the sectors merge gradually into one another: there are no sharp boundaries between them.

It will be noted that in many (but not all) cases the interpretations given are reasonably consistent with the traditional meanings of the houses. So perhaps Sakoian and Acker, and the other traditional astrologers, were not so wrong after all. Nevertheless, the new system is clearly superior to the old. It offers more detail (twenty-four sectors in place of twelve), and it is based on hard empirical evidence.

In fact, even this does not exhaust the complexity of John Addey's findings on the meaning of Mars in the diurnal circle, for he also presents findings[13] on the phases of Mars in the sixth and fifteenth harmonics. But these are not presented in enough detail to be included in the scheme (Figure 11).

John Addey's work was incomplete at the time of his death, and he had not made a full presentation of his findings for any of the planets except Mars. Let us hope that others will continue his work, which promises to be one of the most important foundations of the New Astrology.

CHAPTER 23
THE SEARCH FOR TRUTH

In this final chapter, I would like to consider the question of where this book stands in relation to the search for astrological truth, which is (or should be) the quest upon which all serious astrologers are engaged.

The twentieth century has seen an enormous outpouring of new astrological theories and techniques, which have mostly been added on to the old theories and techniques, rather than replacing them. Thus, the modern astrologer has a great variety of theories and techniques between which he can choose. But *how* can he choose? In 'The need for doubt and the need for wonder' I said:

> Traditional astrology (together with the new techniques which have been added on to it) consists of a vast mass of disconnected assertions, which almost certainly are a mixture of truths, half-truths, blind guesses, and superstitions. And, since these assertions do not relate back to any fundamental and proven laws, it is impossible for the astrologer to sort out the wheat from the chaff, since there are no reliable criteria which he can use. By constantly questioning the assertions of astrology, and by digging deep into the fundamental principles which underlie them, we can eventually hope to build up a new astrology on the ruins of the old: an astrology in which the fundamental laws (or some of them) are known, even though there will always be more to discover about how they operate in particular cases. This new astrology is likely to resemble the old astrology in many ways, but it will not be the same. It is likely to be more beautiful than the old astrology, for it will be closer to truth: it will enable us to hear more clearly the 'music of the spheres'.[1]

The astrologers with whom I am most strongly in disagreement are those who seem not to care about this search for truth. They cheerfully adopt all the new techniques (provided that they find them useful), without questioning the fundamental principles which underlie them, and without asking whether these techniques are consistent with their own view about how the universe works. It is difficult to see how this unselective approach can be justified, since it cannot be possible that *everything* is true. To take four examples: If the Placidus system of house division is valid, then how can Equal house also be valid? If secondary progressions (a day for a year) are valid, how can solar-arc progressions or one-degree progressions also be valid? If composite charts are valid, how can Davison's Relationship Horoscope also be valid? If harmonic charts are valid, how can planetary harmonic charts also be valid? We are of course at liberty to experiment with all these systems, but in the end the search for truth dictates that we must make some hard choices. We must say 'I believe that this one is right and that one is wrong; this one fits in with my model of how the universe works, and that one does not'. If we want to retain both systems, we must at least have some view about how they are consistent with each other and how they can both be fitted into an overall model of truth.

The hard school of scientific astrologers (whose bible is *Recent Advances*[2]) stands at the opposite extreme from the unselective astrologers: the scientific astrologers believe that no technique should be used until it has been empirically and statistically proved to be valid, with all the paraphernalia of control groups, replication studies and the rest. In the hard, pure world of thought which they inhabit, they are doubtless correct: and yet, if their advice was to be followed, all practising astrologers would go out of business while awaiting the scientists' findings (and so would nearly all psychotherapists and psychological counsellors, since the theories of Freud, Jung, Perls, Assagioli, Maslow and the rest are just as unproven and anecdotal as those of astrology). We are searching for truth, but at the same time we have to go on acting in an imperfect world, armed with our present imperfect understanding. If we waited until the complete truth was laid before us, we would all be dead.

So then, if we cannot wait for all the t's to be crossed and the i's to be dotted through scientific experiments, how are we to distinguish between truth and falsehood? How can we select between different astrological theories and techniques, other than on the basis of pure whim and fancy (choosing, perhaps, those techniques which we find enjoyable to use and easy to understand, and rejecting those which we find less enjoyable or more difficult)?

The answer, as I have already suggested, is that we must be guided by the search for a consistent and coherent vision of the nature of reality. And here I must refer to one of the great books of the twentieth century: Michael Polanyi's *Personal Knowledge*. Polanyi says:

> Personal knowledge in science is not made but discovered, and as such it claims to establish contact with reality beyond the clues on which it relies. It commits us, passionately and far beyond our comprehension, to a vision of reality . . . . Such is the true sense of objectivity in science . . . . I called it the discovery of rationality in nature, a name which was meant to say that the kind of order which the discoverer claims to see in nature goes far beyond his understanding: so that his triumph lies precisely in his foreknowledge of a host of yet hidden implications which his discovery will reveal in later days to other eyes.[3]

Elsewhere he says:

> Scientific discovery reveals new knowledge, but the new vision which accompanies it is not knowledge. It is *less* than knowledge, for it is a guess; but it is *more* than knowledge, for it is a foreknowledge of things yet unknown and perhaps inconceivable. Our vision of the general nature of things is our guide for the interpretation of all future experience. Such guidance is indispensable. Theories of the scientific method which try to explain the establishment of scientific truth by any purely objective formal procedure are doomed to failure. Any process of enquiry unguided by intellectual passions would inevitably spread out into a desert of trivialities. Our vision of reality, to which our sense of scientific beauty responds, must suggest to us the kind of questions that it should be reasonable and interesting to explore. It should recommend the kind of conceptions and empirical relations that are intrinsically plausible and which should therefore be upheld, even when some evidence seems to contradict them, and tell us also, on the other hand, what empirical connections to reject as specious, even though there is evidence for them – evidence that we may as yet be unable to account for on any other assumptions. In fact, without a scale of interest and plausibility based on a vision of reality, nothing can be discovered that is of value to science; and only our grasp of scientific beauty, responding to the evidence of our senses, can evoke this vision.[4]

Polanyi shows that all the great advances and discoveries in science have been prompted by the innovating scientist's faithfulness to a vision of reality, his sense of intellectual beauty, and his search for a universal rationality. In former times, from Pythagoras through to and including the times of Copernicus and Kepler, this was openly acknowledged to be the quest upon which scientists were engaged; but more recently the belief has grown that scientists *ought not* to be engaged on such a search for rationality: they should simply be collectors of facts, from which no rationality or beauty can be expected, so that, as Polanyi says, even a telephone directory becomes a scientific theory. Yet Polanyi shows that in fact the quest for a universal rationality has not died, and is still the channel through which our understanding of the world is advanced. For example, he shows that Einstein's formulation of the theory of relativity was *not* the result of his attempt to explain some facts which did not fit in with Newtonian physics, but was the result of his awareness of the *imperfect rationality* of Newton's model of the universe: an awareness which had first come to him at the age of sixteen, before he knew of any scientific experiments which conflicted with Newton's model. Because of this awareness, he was driven to make sense of the universe in a new way.

This search for beauty and rationality is also a search for *meaning*. Facts in themselves can be memorized, but they cannot be *understood* unless we give them a *meaning* (or an *interpretation*) which relates them to our overall vision of reality. Of course, we may sometimes be obliged to revise our vision in the light of newly-discovered facts (as, for example, Gauquelin did when he found that there was some truth in astrology), but we must have a vision which goes beyond the proven facts and attempts to make sense of them within an overall framework. Nowadays, many scientists do not state what their models of the universe (or visions of reality) are; but they must be there, if they are to make any sense of their own findings.

Astrological models of reality

All this is of the greatest relevance to astrology, for astrologers are necessarily working from a vision of reality which goes beyond Einstein, beyond Jung, beyond Darwin, and beyond the great religious thinkers; they are concerned with a different dimension of reality, for which no accepted model exists. It is, therefore, of the greatest importance that astrologers should clarify – for themselves and for others – the models of the universe from which they are working, and that, in their search for astrological truth, they should make clear what kind of ultimate rationality and what kind of universal beauty they expect to find.

If we look at astrological theory and practice from this point of view, we find that different astrologers are working (or at least appear to be working) from different models of reality. Some of these models are, to some extent, compatible with one another: that is to say, it would be possible to fit them together with other models into a single overall model. But in other cases the models seem to be incompatible with one another, so that if we are concerned with astrological truth, we will have to adhere to one model and reject another.

In this chapter I would like to look at some of these basic models (or visions of reality) from which different astrologers appear to be working. To do this fully would require a whole book on its own. I will be dealing only with *natal* astrology, and so will exclude both mundane astrology and predictive astrology; also, I will not be looking at models of ultimate causes, but will be dealing only with models of the *nature of astrological effects within the birth chart*.

First, there is what we can call the **boxes** model. According to this model, the heavens are divided into a number of separate 'boxes' with sharp (or sharpish) boundary-lines between them. There are two such systems of boxes: one, called the 'signs', divides the zodiacal circle; the other, called the 'houses', divides the diurnal circle. (The word 'houses' is, of course, an indication that we are dealing with a 'boxes' model: houses, like boxes, have sharp boundaries and solid walls, so that one can be inside them or outside them but one cannot, except at a narrow point called the entrance or threshold, be half in and half out.) Each of the boxes has a particular set of qualities attached to it. The planets, as they move round the zodiacal and diurnal circles, move in and out of the various boxes, and take into themselves the qualities of the particular boxes which they are in at the moment. Thus, the influence which a planet exerts upon mankind at a particular moment is a blend of the permanent qualities of the planet with the qualities of the boxes which it temporarily occupies. But also, the two sets of boxes are always moving in relation to each other (causing, for instance, different signs to appear on the Ascendant), and this has its own effect, even when no planets are involved.

Traditionally there are twelve signs and twelve houses, and

so the boxes model assigns a special importance to the number Twelve (and also to the numbers Three and Four of which it is composed, since the boxes are divided into three groups of four and also four groups of three). Other numbers such as Five and Seven, into which Twelve is not divisible, can play no part in the scheme. (It is true that we talk of the fifth house and the seventh house, but these figures refer only to numbers of twelfths: the fifth house is five-twelfths of the way round the circle from the Ascendant. This is the only way in which numbers such as Five can be admitted into the boxes model.)

Traditionally also, the two sets of boxes are related to each other. The signs 'start' with Aries, and the houses 'start' at the Ascendant; and Aries has qualities in common with the first house from the Ascendant, and so on round the circle. The two sets of boxes are in phase with each other only when the first point of Aries (which the Sun reaches at the spring equinox, when it is half-way from furthest south to furthest north) is on the Ascendant (where it is half-way from furthest below the horizon to furthest above). Zipporah Dobyns[5] has perhaps gone further than any other astrologer to convert this into a coherent theory, claiming that the twelve houses and signs each represent the twelve 'letters' of the astrological alphabet.

The boxes model is of course the most time-honoured system in astrology, and is still very much alive. Astrologers have been reinterpreting the boxes in the light of modern psychological theories and the conditions of the modern world, and many of them are doing excellent work within this framework. Even though the boxes theory may not be the whole truth, it still seems to be a very serviceable approximation to it, and so we should perhaps be wary of abandoning it. And yet I for one have always been intellectually dissatisfied with the boxes model: it seems to be inherently implausible that there should be these sharp dividing-lines in the sky, and that the number of boxes should be unalterably twelve. If we accept Polanyi's version of the nature of scientific enquiry, then such intellectual doubts must play their part in forming our vision of reality.

Secondly, there is the **networks** (or **linkages**) model, according to which the qualities of a planet at a particular time are determined by the links it has with other planets (which planets it is connected with, and by what kind of link). The theory of astrological *aspects* is built up from such a model. Essentially the theory says that, when the angular relationship between two planets comes close to a specific fraction (e.g. one-third of the circle), then a link is formed between them which has the quality of the number which is the divisor of the fraction (in this case, the number Three). Hence the planets are connected together by a complex pattern of links, which together comprise a *network* of inter-planetary relationships. Traditionally the model has been concerned only with One, Two and Three (and multiples of these numbers), but there is no reason why it cannot be expanded to include every other number. Hence the networks model (unlike the boxes model) does not give a special significance to any particular number.

In Western astrology, the networks model has traditionally existed alongside the boxes model, and modern astrologers are mostly very adept at combining the two theories. But these two theories are very uneasy bedfellows. According to the boxes model, provided that a planet is within a box, it does not matter *where* it is in the box: thus 1° Taurus has the same meaning as 29° Taurus. But, according to the networks theory, a planet at 0° Aquarius is *square* to a planet at 1° Taurus (and also to a planet at 29° Aries), but is *trine* to a planet at 29° Taurus (and also to a planet at 1° Gemini). Hence the networks theory seems to be saying that there are no sharp dividing-lines between boxes, and that 29° Taurus is more similar to 1° Gemini than to 1° Taurus, and this is tantamount to saying that there are no boxes at all. Margaret Hone's theory of dissociate aspects[6] (according to which a trine from 0° Aquarius to 29° Taurus is *weaker* than a trine from 0° Aquarius to 1° Gemini, because it crosses the boundary between signs) is an attempt to deal with this problem, but it does not dispel the essential contradiction between the two theories.

Thirdly, there is the **waves** model, which is particularly associated with John Addey and therefore with harmonics. The theory of waves is in direct contradiction to the theory of boxes, because it claims that there are no sharp dividing-lines in the sky; rather, there is a gradual movement from one quality to another, or a gradual oscillation between different qualities.

The waves model has been fully set out by John Addey in Chapters 2 and 3 of *Harmonics in Astrology*. He explains that: 'The kind of waves we shall be dealing with are called sine waves . . . . Sine waves are just those which are produced by simple harmonic motion such as is produced by a swinging pendulum or a tuning fork or the motion of light waves'.[7] The first harmonic consists of a single wave with a length of 360 degrees; the second harmonic consists of two waves, each with a length of 180 degrees; and so on. These waves can be drawn around the circle. Thus, the diagram on the front cover of this book shows concentric harmonics, from the first harmonic with only one wave to the ninth with nine waves.

According to wave theory, then, the qualities of a planet at a particular time are determined by its position on a number of harmonic waves. In one harmonic, it may be at the crest of the wave; in another harmonic it may be at the trough; and in a third harmonic it may be half-way between the crest and the trough.

This is the best model for explaining Gauquelin's finding (which we discussed in the first part of Chapter 22) that the planets have particular 'zones of high intensity' within the diurnal circle. In fact, Gauquelin's finding that the peaks of intensity near the Ascendant and the M.C. are stronger than those near the Descendant and the I.C. can be explained by a combination of a strong fourth-harmonic wave with a weaker third-harmonic wave. The peaks near the Ascendant and M.C. are strong because, in these areas, the crests of the H4 and H3 waves nearly coincide; but the peaks near the Descendant and I.C. are weaker because here the crest of the H4 wave nearly coincides with the trough of the H3 wave.

In *Harmonics in Astrology,* John Addey claims that such waves exist not only in the diurnal circle (replacing the traditional houses) but also in the zodiacal circle (replacing the signs) and in the aspectual circle (replacing the traditional aspects). He suggests that the whole of astrology can be explained in terms of harmonic waves. He is thus putting forward a clear statement of a 'vision of reality' which has revolutionary implications for what is true, and what is not true, within astrology.

But in fact, when we turn to John Addey's later research on the meaning of Mars in the diurnal circle (which we examined in the second part of Chapter 22), we find that the waves model becomes inadequate. For wave theory suggests that everything can be explained in terms of crests and troughs: that is, the presence or absence of a particular quality. But Addey's later research suggests that within each harmonic there are a number of different qualities which are represented by different points around the circle, so that it is not simply a question of the presence or absence of one quality. We can see clearly that this is true within the first harmonic (which is the circle of the heavens themselves): North is the opposite of South, but also East is the opposite of West, and North-East is the opposite of South-West. Therefore we cannot speak of the first harmonic in terms simply of the presence or absence of Northness, since this would imply that there was no difference between Eastness and Westness. But Addey's research shows that (as we would expect) this is true in the other harmonics also. Thus the second harmonic is that in which North/South is the opposite of East/West, and also N.E./S.W. is the opposite of N.W./S.E. In the same way, Mars in the second, third and fourth harmonics has a number of different phases, each of which has a distinct meaning.

In order to accommodate this, we have to abandon the waves model and to develop instead a **cycles** model, in which the harmonics are conceived of as cycles rather than as waves. Thus, the first harmonic of the zodiacal circle is seen as a single cycle of 360 degrees, within which 0° Aries is the opposite of 0° Libra, 0° Taurus is the opposite of 0° Scorpio, and so on. The second harmonic consists of two cycles, each of 180 degrees: 0° Aries is the same as 0° Libra, and both are opposite to 0° Cancer and 0° Capricorn. The third harmonic consists of three cycles, and so on. Within this model, we do not seek to discover the positions of the crests and troughs; rather, we seek to discover the qualities attached to each phase of each cycle. I believe that this is the model which best explains John Addey's findings, and that in his late work Addey can be regarded as having claimed that the whole of astrology can be explained in terms of, not waves, but cycles.

According to the cycles model, then, the qualities of a planet are determined by its position within a large (in fact, a potentially infinite) number of harmonic cycles. Each of the cycles – and each phase within each cycle – has to be studied separately before we can understand the total picture. In fact, one of the problems with the cycles model – and also with the networks model when it is extended beyond the traditional aspects – is that it enormously increases the complexity of astrology, and makes it clear that the total picture can *never* be grasped. The simplicity of the boxes model – in which, once we have understood the meaning of the twelve boxes, we have understood it all – has vanished. For some people, this would be a good reason for rejecting the cycles model. But the *concept* of harmonic cycles is a very simple one, and seems to me to satisfy Polanyi's criteria of rationality and intellectual beauty.

There are two other models which we should perhaps mention. The first of these is the **structures** model which underlies Ebertin's theory of midpoints.[8] Ebertin and his followers believe that the qualities of a planet at a particular time are partly determined by whether it is at the midpoint (or in aspect to the midpoint) between two other planets or Angles. Within any birth chart a large number of these midpoint relationships exist, so that complex structures of interplanetary relationships are built up. The meaning of each planet is constructed from the meanings of all the midpoints which it occupies. This can be seen as an elaboration of the networks model, allowing linkages between three as well as between two planets; but it greatly complicates the networks model, and makes it far more difficult to understand at the conceptual level. The question of the compatibility between the structures model and the other models is a complex subject which I do not wish to go into here; but from a practical point of view, both midpoints and harmonics introduce vast new areas of complexity and uncertainty, and it is difficult to develop in both of these directions at the same time. Most practising astrologers will therefore want to choose between them. (In Ebertin's theory, the fourth harmonic is regarded as supremely important, and all other harmonics (except the first) are ignored. If we were to introduce other harmonics into midpoint theory, the number of midpoints would increase astronomically and would in effect become unmanageable.)

Finally we can mention the model which seems to underlie the theory of planetary harmonics, which we discussed in Chapter 21. I like to think of this as the **Monopoly** model, because in the game of Monopoly a player receives £200 every time he passes Go: in the same way, according to the planetary harmonics theory, a planet receives a sudden input of energy every time it passes 0° Aries. The same type of model perhaps underlies the old theory of 'angular, succedent and cadent houses', according to which a planet receives a sudden input of vigour each time it passes one of the Angles. This model seems to be incompatible with all the other models.

A personal view

Having set out (at a very basic level) a number of models of the nature of astrological effects, I must now state my own position. In doing so I must emphasize that I am not saying that my views are necessarily correct: I am saying only that they appear to me to be true at the present time, in the light of my understanding of the facts and my efforts to make sense of them. Other astrologers, seeing the facts from a different viewpoint and perhaps with a different sense of intellectual rightness, may reach different conclusions. And yet at the same time I *believe* my views to be true, since they are the only way in which I can make sense of astrology. They constitute *personal knowledge* in Polanyi's sense of the term .

So far as the zodiacal and diurnal circles are concerned, it seems to me that the cycles model must be correct. It corresponds to everything we know about the physical universe, and in particular about the solar system: the circular shape of the planets, the circular nature of their spin and the near-circular nature of their orbits, the cycles of day and night and of the seasons, the cycle of the Great Year. These are continuous cycles which have no beginning and no ending, and within which there are no sharp dividing-lines between stages or phases. It is reasonable to suppose that the cycles of astrology, which are based on these physical cycles, are similar to them. We may find it convenient to regard the zodiac as starting at 0° Aries, and the diurnal circle as starting at the Ascendant (just as we find it convenient to celebrate the New Year on January 1st), but we should know that really these starting-points and finishing-points are man-made, and that in nature there is no fixed starting-point, but only a continuous cycle: 'All the rivers flow to the sea, yet the sea is not full: unto the place from whence the rivers came, thither they return again'.[9]

And similarly we should know that there are no sharp boundaries within the cycle. June may be a warmer month than May, but the weather does not suddenly become warmer on June 1st. Similarly, if we believe that there are qualities attached to the section of the zodiac which we call Aries, and other qualities attached to the section which we call Taurus, we must believe that these qualities merge gradually into one another.

These conclusions seem to me to be self-evident for the first harmonic of the zodiacal circle (which is the cycle of the year) and for the first harmonic of the diurnal circle (which is the cycle of the day and night). But if so, it follows that they must be true also for the other harmonics. The third harmonic of the zodiacal circle is the cycle from Fire through Earth, Air, Water and back to Fire: and if Aries merges gradually into Taurus, it follows that, in the third harmonic, Fire merges gradually into Earth.

This fits in with John Addey's findings on the meaning of Mars in the diurnal circle. The nature of Mars is affected by its positions on a number of harmonic cycles, and each phase of each cycle has its own specific meaning. And yet, when we examined John

Addey's findings in Chapter 22, we explained them in terms of a number of boxes which we called sectors: without creating these boxes it would have been difficult to comprehend the meaning of the cycles. This points to the universal tendency of the human mind to categorize in order to comprehend. We split off May from June, spring from summer, Aries from Taurus, even when we know that really there are no sharp boundaries between them. So long as we remember that the creation of these boxes is only for convenience and ease of understanding, and that in reality there is a gradual transition from the qualities of one box to those of the next, then no harm is done. But there is always the danger that the boxes will solidify, and will come to seem to have their own separate existence, so that we will cease to believe in a continuous cycle and believe instead in the boxes which we have created.

In the case of the other nine planets in the diurnal circle, and of all ten planets in the zodiacal circle, there is as yet no system of interpretation which can take over from the traditional system of twelve signs and twelve houses. We must therefore either use the traditional system or abandon altogether the attempt at interpretation. Even if we believe in the cycles model, we may still use the traditional system because we believe it to be an acceptable approximation to reality; but we must remember that the boxes are man-made and that in reality there must be a gradual transition from one phase of the cycle to the next.

So far I have been talking entirely about the diurnal and zodiacal circles. But when we turn to the *aspectual* circle, we have an entirely different situation.

The aspectual circle is the cycle of aspects between any pair of planets, from conjunction to opposition and back to conjunction. The best-known aspectual circle is the cycle of the phases of the Moon, which is the Sun-Moon aspectual circle. If we think about this cycle, we realize that it does have a clear and obvious starting-point, which is the conjunction (the New Moon). The cycle of the relationship between the two bodies starts at the point where they are in union, or Oneness, with each other, and it ends at the point where this Oneness is again achieved.

Not only is there a clear starting-point and finishing-point, but also there are clearly defined and meaningful points within the cycle. Half-way round, there is a point (the opposition, or Full Moon) where the relationship between the two bodies clearly has the nature of Twoness; and within the cycle there are two points (the trines) where the relationship has the nature of Threeness; and so on. In other words, at these points *links* are formed between the planets which have the nature of a particular number. At these points there is no gradual transition, but the sudden *peaking* of a particular quality.

Therefore it seems to me that the most appropriate model for the aspectual circle is the networks model rather than the cycles model. In saying this I am of course following astrological tradition, but extending it to include aspects based on numbers such as Five, Seven and Nine which astrologers have traditionally ignored. I am suggesting that the chart contains a complex network of interplanetary links based on the aspects between pairs of planets, and that this network can be studied in detail through the use of harmonic charts. Moreover it seems to me that we know far more about how to interpret these networks of aspects than we know about how to interpret the cycles of the diurnal and zodiacal circles. Therefore the exploration of networks of aspects has been the main subject of this book.

The way in which we interpret aspects between planets will be different, depending on whether we are using the networks model or the cycles model. Within the networks model, we would expect to find sudden and sharp peaks at the points where the links between planets are strongest. But within the cycles model we would expect to find a continually flowing pattern of gradual change, in which the conjunction was not necessarily more important than any other point in the cycle, and in which 'separating' and 'applying' aspects had different meanings because they occupied different phases of the cycle. My own adherence to the networks model is based on the fact that throughout the studies of harmonic charts which led to the writing of this book, the one point which has most impressed me is that *close aspects are extremely strong.* When an aspect between two planets in a chart is very close to exactitude, it seems that a vey powerful link is formed between these planets, which often dominates the personality. On either side of the point of exactitude, the strength of the aspect fades away, although a less-exact aspect may still be strong if it forms part of a network of aspects involving three or more aspects. This seems to be entirely consistent with the networks model, but it would be difficult to explain by means of the cycles model.

I am aware, however, that in adhering to the networks model for explaining the aspectual circle, I am departing from John Addey's belief that the whole of astrology can be explained in terms of waves or cycles. Addey himself seems to be somewhat inconsistent in this respect. In Chapter 14 of *Harmonics in Astrology* he explains aspects in terms of the networks model, but elsewhere[10] he discusses the aspectual circle in terms of waves or cycles and presents findings which seem to be more consistent with a cycles model. It is possible that the real truth about the aspectual circle lies in a complex model which contains elements both of cycles and of networks, but I have not yet been able to conceive how such a model would work.

Celestial latitude and lunar parallax

Here I would like to mention two technical problems: celestial latitude[11] and lunar parallax.[12] They are very inconvenient problems, and many astrologers would prefer to ignore them; but we cannot afford to ignore them if we wish to face up to astrological reality.

First, **celestial latitude.** When we measure the aspects between planets, we are measuring the angular relationship between their positions on the zodiacal circle, which is the circle of the ecliptic: and, if we are interested in exact aspects in harmonic charts, we have to define the planets' positions on the ecliptic with very great exactitude. But here the problem arises that (with the exception of the Sun) the planets are very rarely exactly on the ecliptic. They have celestial latitude, which means that they lie some distance away from the Sun's path. Pluto's latitude may be as much as 17 degrees, and some of the other planets may have a latitude approaching 8 degrees. The positions on the ecliptic, which we define with such exactness, are not the actual positions of the planets but the points on the ecliptic which are closest to the planets. Therefore the true angular relationship between the planets themselves may be different from what we have measured. We are measuring not the aspects between the planets themselves, but the aspects between their nearest points on the ecliptic.

It would be possible (though difficult) to correct the angular relationship between each pair of planets to allow for celestial latitude. To do this, we would have to define the Great Circle connecting each pair of planets (since we would no longer be relying on the single Great Circle of the ecliptic), and to measure the angular relationship within this Great Circle. But, by doing this, we would invalidate all the interpretations given in this book, and also to a large extent the findings of other astrologers, which are based on precise calculation of zodiacal longitude with no allowance for latitude. Therefore, unless we want to wipe the slate clean and start again, we have to believe that the important influencing points are not the positions of the planets themselves, but their nearest points on the ecliptic. The circle of the ecliptic becomes, as it were, a 'ring of power' around the Earth, through which all astrological effects are channelled.

But, if this is the case, then it seems that it *must* be true that the Gauquelin system of measurement along the diurnal circle (which we described in Chapter 22) is the wrong system, since it is based on the times when the *planets themselves* rise and culminate, rather than on the times when their nearest points on the ecliptic rise and culminate. We must, in other words, revert to a system such as Placidus or Equal house which measures positions along the ecliptic itself.

Second, **lunar parallax.** The problem here is that if we calculate the Moon's longitude from the ephemeris, we obtain its position *relative to the centre of the Earth*. If we wish to calculate its position as seen from a specific point on the surface of the Earth, we have to correct for parallax, and this correction can be as great as 1 degree (which of course is magnified to 9 degrees in the H9 chart). Thus it might seem necessary to correct for parallax in every chart, in order to obtain the Moon's position as seen from the actual birthplace. However, it seems that most astrologers who have researched into the parallax question have come to the conclusion that better results are obtained from *not* correcting for parallax.[13]

But if we decide not to correct for parallax, we must again consider the implications of this for our model of how astrology works. By not correcting for parallax, we are measuring the positions of the planets *as 'seen' from the centre of the Earth*. The implication is that the planets do not directly affect the person who is standing on the Earth's surface, but *affect the Earth itself*: the person is affected only because he is part of the 'collective consciousness' of Earth. Elsewhere[14] I have suggested that there is a Moon-mind, an Earth-mind, and so on, and that astrology is the means by which these minds communicate with one another. I do not wish to develop this argument here, except to say that astrologers should be thinking about these issues in their efforts to make sense of astrology.

These problems of latitude and parallax give rise to a number of difficult questions, for which no easy answers are available. But we must be trying to answer them if we are working towards a valid model of astrological reality. Until we have solved them, our model of reality is tentative and incomplete, and may in some respects be self-contradictory.

The meaning of Number

I have deliberately left to the end the question of the meaning of Number, which is perhaps the most important (and certainly for me the most exciting) issue discussed in this book. I have tried (especially in Chapter 20) to develop a scheme for the interpretation of Number which is logically coherent as well as consistent with empirical evidence, and it is necessary to consider the philosophy underlying such a scheme.

Polanyi makes it clear that among the ancient Greeks there were two contrasting schools of scientific thought. The first was the Ionian school, which believed in simply listing, counting, describing and categorizing the things which happened to exist. The other was the school of Pythagoras. Polanyi says:[15]

> Pythagoras and his followers did not, like the Ionians, try to describe the universe in terms of certain material elements (fire, air, water, etc) but interpreted it exclusively in terms of numbers. They took numbers to be the ultimate substance, as well as the form, of things and processes. When sounding an octave they believed they could hear the simple numerical ratio of 1:2 in the harmonious chiming of the sounds from two wires whose lengths had the ratio 1:2. Acoustics made the perfection of simple numerical relations audible to their ear. They turned their eyes towards the heavens and saw the perfect circle of the sun and moon; they watched the diurnal rotation of the firmament and, studying the planets, saw them governed by a complex system of steady circular motions; and they apprehended these celestial perfections in the way one listens to a pure musical interval. They listened to the music of the spheres in a state of mystic communion.

Polanyi goes on to say:

> The revival of astronomical theory by Copernicus after two millennia was a conscious return to the Pythagorean tradition . . . . After Copernicus, Kepler continued wholeheartedly the Pythagorean quest for harmonious numbers and geometrical excellence . . . . He even went so far as to write down the tune of each planet in musical notation.

However, after Kepler (and after Galileo who stood midway between the two schools), Polanyi reports that there was:

> . . . a definite change from the Pythagorean to the Ionian conception of theoretical knowledge. Numbers and geometrical forms are no longer assumed to be inherent as such in Nature. Theory no longer reveals perfection; it no longer contemplates the harmonies of Creation. In Newtonian mechanics the formulae governing the mechanical substratum of the universe were differential equations, containing no numerical rules and exhibiting no geometrical symmetry. Henceforth 'pure' mathematics, formerly the key to nature's mysteries, became strictly separated from the *application* of mathematics to the formulation of empirical laws . . . . Mathematics represented all rational thinking which appeared necessarily true; while reality was summed up in the events of the world which were seen as contingent – that is, merely such as happened to be the case.

Although I have dissented in a small way from some of John Addey's views, I am his committed follower in wanting to return to the Pythagorean tradition in astrology. Astrology itself is not necessarily Pythagorean: there is plenty of Ionian astrology, concerned simply with listing and categorizing what has been found to be true. But harmonic astrology is Pythagorean, because it is concerned with the nature of Number and believes Number to be the key (or at least one of the keys) to the mysteries of the universe.

In this book I have tried to set out a theory of the meaning of Number which is consistent within itself and adds up to a possible model of this aspect of reality. The important thing to note about this system is that it is *multiplicative*: that is to say, if a number is the product of two other numbers, then it contains the qualities of these two numbers combined. For instance, once we have established that Four has the meaning of 'striving towards . . .', then the theory tells us that forty-four means 'striving towards Elevenness': we know this to be the case even before we know what Elevenness means.

If the system is multiplicative, it cannot also be *additive*: that is to say, if a number is the *sum* of two other numbers, it does *not* contain the qualities of those two numbers combined. Some astrologers have sought the meaning of the number Seven in the fact that seven equals four plus three, so that it must combine Fourness with Threeness. But seven is also five plus two, and also six plus one: so there is no reason why we should regard four and three as being particularly connected with seven. In fact, in an additive system, *every* number would contain the qualities of *every* preceding number. In a multiplicative system, however, Seven is a prime number (made up of no numbers except itself and One), and so has a new quality which is not contained in any preceding number. The number which combines Fourness with Threeness is not Seven, but Twelve.

Also, if the system is multiplicative, it cannot be *digital.* We cannot say that a number combines the qualities of its digits (for instance, that 27 combines Twoness with Sevenness), for these digits are the result of the man-made decimal system and are not an inherent property of the number. In a duodecimal system,[16] which might seem more suited to astrology than a decimal system, 27 would be written as 23, but it would still be the same number: also, 27 is still the same number if we write it XXVII. If we are thinking of the quality of the number Twenty-seven (the number itself, rather than the names we give it or the way in which we write it), the important thing about it is that it is $3 \times 3 \times 3$.

I have been discussing the theory of Number as an abstract theory. Nevertheless, Pythagorean science, like Ionian science, must of course be based on observation and be consistent with the observed facts. I have deliberately refrained from a study of traditional numerology, because I wanted to ensure that my conclusions on the nature of Fourness, Fiveness, Sevenness and Nineness were based on what I found within actual astrological charts, and I would prefer to leave it to others to decide whether these findings are consistent with the traditions of numerology. This is because I fear that, since the demise of Pythagoreanism as a respectable approach to science, numerology may have become a purely 'occult' and mystic subject which has lost contact with reality. The aim of Pythagoreanism is to unite mysticism with reality, rather than to promote mysticism for its own sake.

However, if it is decided that the statements of John Addey, myself and others about (for instance) the nature of Fiveness have any validity, then it follows (within the Pythagorean tradition) that these are properties of the number Five itself, rather than simply of the fifth astrological harmonic. That is to say, they can be expected to apply also to all kinds of things and processes which have the quality of Fiveness: for instance, five-sided figures, five-petalled flowers, music with five beats to a bar, poems with five lines, processes with five stages, groups of five people. All of these are entities with five parts, so that the relationship between any two of the parts has the nature of a quintile. By studying all of these entities, and by interpreting from theory to reality and back again to theory, we could hope to get closer and closer to the quintessential nature of Fiveness. And by relating this to our understanding of the quintessential nature of other numbers, we would come closer to unlocking the secrets of the universe, so far as that is possible within the limits of man's understanding.

Of course, it *may* be true that this is a quest for the non-existent: it *may* be true that there is no pure and ultimate Fiveness, and that there is no ultimate rationality governing the universe. The universe may turn out to be simply a gigantic telephone directory, full of disconnected facts. But the quest for an ultimate rationality and beauty (like the quest for God, with which it is connected) seems to be a vital quest for man: he needs to engage in it in order to make sense of the world in which he finds himself.

It seems to me that the two greatest quests upon which men and women are engaged are, firstly, this quest for an understanding of themselves in relation to the total universe and its causes ('Thou shalt love the Lord thy God . . .'), and, secondly, the quest for better relationships with (through better understanding of) their fellow men and women ('. . . and thou shalt love thy neighbour as thyself'). A large part of this book has been concerned with the second of these quests: it has been concerned with improving our understanding of the individual birth chart so that we can better understand, and better relate to, our fellow men and women. But astrology offers the opportunity to relate the two quests to each other. We can learn about the Ultimate from studying individual birth charts, and then we can use our understanding of the Ultimate for the purpose of understanding our fellow men. We can see heaven, if not in a grain of sand, at least in the unfolding of an individual life. Astrology helps us to see man as part of a universal process and as governed by universal laws; and, while some see this as belittling man by making him subject to forces greater than himself, I prefer to see it as magnifying him by making him part of a larger scheme.

I would like to end this book, not with my own words, but with those of Olaf Stapledon from his 'myth of the future' *Last and First Men.*[17] These words were not written in an astrological context, but they seem to acquire a new meaning from being placed in such a context; and they have a particular relevance at a time when man, through misunderstanding both his relationship to the universe and his relationship with his fellow men, may be about to destroy himself with his own technology.

> Great, and terrible, and very beautiful is the Whole; and for man the best is that the Whole should use him.
>
> But does it really use him? Is the beauty of the Whole really enhanced by our agony? And is the Whole really beautiful? And what is beauty? Throughout all his existence man has been striving to hear the music of the spheres, and has seemed to himself once and again to catch some phrase of it, or even a hint of the whole form of it. Yet he can never be sure that he has truly heard it, nor even that there is any such perfect music at all to be heard. Inevitably so, for if it exists, it is not for him in his littleness.
>
> But one thing is certain. Man himself, at the very least, is music, a brave theme that makes music also of its vast accompaniment, its matrix of storms and stars. Man himself in his degree is eternally a beauty in the eternal form of things. It is very good to have been man.

APPENDIX I

METHOD FOR CALCULATING HARMONIC POSITIONS BY HAND

The steps to be followed are:

1 Convert the planet's zodiacal position into absolute longitude, using the table in Appendix II.
2 Multiply the absolute longitude by the harmonic number (e.g. 4 for the fourth harmonic).
3 If the resulting figure is greater than 360, subtract 360. Continue to subtract 360 until you have a figure which is less than 360.
4 Convert back into zodiacal longitude.

We can give two examples:

(a) Find the fifth-harmonic position of a planet at 17°21′ Gemini.

$$17°21' \text{ Gemini} = 60° + 17°21' = 77°21'$$

$$77° \times 5 = 385° \qquad 21' \times 5 = 105' = 1°45'$$

$$385° + 1°45' = 386°45'$$

$$-1 \times 360 = 360$$

$$26°45' = \text{Aries } 26°45'$$

(b) Find the seventh-harmonic position of a planet at 10°11′ Pisces.

$$10°11' \text{ Pisces} = 330° + 10°11' = 340°11'$$

$$\begin{array}{r} 340° \\ \times\ 7 \\ \hline 2380° \end{array} \qquad \begin{array}{r} 11' \\ \times 7 \\ \hline 77' \end{array} = 1°17'$$

$$\begin{array}{r} 2380° \\ +\quad 1°17' \\ \hline 2381°17' \end{array}$$

$$\begin{array}{r} -6 \times 360 = \underline{2160} \\ 221°17' = \text{Scorpio } 11°17' \end{array}$$

APPENDIX II

TABLE OF ABSOLUTE LONGITUDE

0° Aries	= 0°	0° Libra	= 180°
0° Taurus	= 30°	0° Scorpio	= 210°
0° Gemini	= 60°	0° Sagittarius	= 240°
0° Cancer	= 90°	0° Capricorn	= 270°
0° Leo	=120°	0° Aquarius	= 300°
0° Virgo	=150°	0° Pisces	= 330°

Thus 16° Leo = 120° + 16° = 136°

APPENDIX III
HARMONIC ORBS

The recommended orbs for aspects between planets, for use in both radical and harmonic charts, are as follows (reproduced from Chapter 3):

	Ordinary aspects	*'Close' aspects*
Conjunction	12°	2°
Opposition	6°	1°
Trine	4°	40′
Square	3°	30′
Sextile	2°	20′
Semi-square	1°30′	15′
Semi-sextile } Quincunx }	1°	10′

The recommended orbs for use in *synastry* (harmonic matrices) and in *composite charts* are different from these, and are presented in Chapter 17.

The recommended orbs for use in *harmonic transits* are presented in Chapter 19.

APPENDIX IV

DATA AND SOURCES FOR CHARTS USED IN THE TEXT

Abbreviations

ABC Lois M. Rodden, *The American Book of Charts,* San Diego: Astro Computing Services, 1980.
FCN Jadwiga M. Harrison, *Fowler's Compendium of Nativities* (with notes on sources by Charles Harvey and the A.A. Data Section), Romford: L.N. Fowler & Co., 1980.
GAU Gauquelin (from his lists of data from birth certificates, unless otherwise stated).
GBR *The Guinness Book of Records,* Edition 24, 1977.
POW Lois M. Rodden, *Profiles of Women,* American Federation of Astrologers, 1979.

GMT Greenwich Mean Time
LMT Local Mean Time
Berne time LMT for Berne (7 E 28)
BST = MET Zone 1h E
BDST = EET Zone 2h E
CDT = CWT EST Zone 5h W
CST Zone 6h W
PDT Zone 7h W
PST Zone 8h W

Birth charts

John ADDEY 15 June 1920, 8.15 a.m. BST, Barnsley (52 N 33. 1 W 29) (From native)
Ludwig van BEETHOVEN 16 December 1770, 1.30 p.m. LMT, Bonn (50 N 43. 7 E 06) (Paul Schmidt *Kosmobiologie* March 1934, via Charles Harvey. The doubts about the authenticity of this chart are discussed in Chapter 14.)
Hector BERLIOZ 11 December 1803, 5.00 p.m. LMT, La Côte-Saint-André (45 N 24. 5 E 15) (GAU)

St BERNADETTE of LOURDES) 7 January 1844, 2.00 p.m. LMT, Lourdes (43 N 07. 0 W 03) (POW: thought to be from birth record.)

Ian BRADY 2 January 1938, 12.40 p.m. GMT, Glasgow (55 N 51. 4 W 16) (From birth certificate, via FCN; also Chester Kemp *Astrological Journal* Spring 1973; but ABC states 12.40 a.m., from biography. The Ascendant for 12.40 p.m. is 16° Taurus, not 12° Taurus as stated in FCN.)

Georges BRAQUE 9 May 1882, 2.30 a.m. LMT, Argenteuil (48 N 57. 2 E 14) (GAU)

Rupert BROOKE 3 August 1887, 7.30 a.m. GMT, Rugby (52 N 23. 1 W 16) (From biography by Christopher Hassall. FCN wrongly states that Hassall says 7.30 p.m.)

Edgar CAYCE 18 March 1877, 3.00 p.m. LMT, Hopkinsville, Kentucky (36 N 52. 87 W 26) (From native's grandmother, via biography and FCN. This time is preferred to the time of 1.30 p.m. which Cayce 'after investigation' – and influenced by an astrologer – decided was the true birth time.)

Winston CHURCHILL 30 November 1874, 1.30 a.m. LMT, Woodstock (51 N 45. 1 W 15) (ABC, from *Jennie* by R.G. Martin.)

Jean COCTEAU 5 July 1889, 1.00 a.m. LMT, Maisons-Laffite (48 N 57. 2 E 08) (GAU)

Camille COROT 16 July 1796, 1.30 a.m. LMT, Paris (48 N 50. 2 E 20) (GAU)

Salvador DALI 11 May 1904, 8.45 a.m. GMT, Cadaques, Gerona (approx. 42 N 11. 2 E 30) (M.E. Jones: confirmed by GAU, *Spheres of Destiny,* who says 'a little before 9.00'.)

Franchet D'ESPEREY 25 May 1856 10.30 p.m. LMT, Algeria (assumed to be Algiers: 36 N 42. 3 E 08) (GAU)

Charles DICKENS 7-8 February 1812, midnight LMT, Portsmouth (50 N 48. 1 W 06) (Langton, *Childhood and Youth of Charles Dickens,* via FCN: through ABC says this is 'dirty data'.)

Eleonora DUSE 3 October 1858, 2.00 a.m. LMT, Vigevano (45 N 18. 8 E 50) (GAU. The positions given in POW are incorrect, apparently due to miscalculation of sidereal time.)

Albert EINSTEIN 14 March 1879, 11.30 a.m. LMT, Ulm (48 N 23. 10 E 00) (GAU: though GAU in *Cosmic Influences on Human Behaviour* wrongly states 1897.)

James ENSOR 13 April 1860, 4.30 a.m. LMT, Ostend (51 N 14. 2 E 55) (GAU)

Maurice ESCANDE 14 November 1892, 4.00 a.m. LMT, Paris (48 N 50. 2 E 20) (GAU)

Bobby FISCHER 9 March 1943, 2.39 p.m. CWT, Chicago (41 N 52. 87 W 39) (From biography, via ABC.)

Scott FITZGERALD 24 September 1896, 3.30 p.m. LMT, St Paul, Minnesota (44 N 57. 93 W 06) (Sara Mayfield, *Exiles from Paradise,* via ABC.)

Zelda FITZGERALD 24 July 1900, 5.33 a.m. CST, Montgomery, Alabama (32 N 23. 86 W 19) (Rectified by R.H. Oliver from time of 5.40 a.m. in family bible: via POW.)

Greta GARBO 18 September 1905, 9.00 p.m. MET, Stockholm (59 N 20. 18 E 03) ('C.C. Zain from her passport and a personal friend', via POW.)

Paul GAUGUIN 8 June 1848, 10.00 a.m. LMT, Paris (48 N 50. 2 E 20) (GAU)

Uri GELLER 20 December 1946, 2.00 a.m. EET, Tel Aviv (32 N 02. 34 E 49) (From native according to ABC; from biography according to FCN.)

Ernest HEMINGWAY 21 July 1899, 8.00 a.m. CST, Oak Park, Illinois (41 N 53. 87 W 47) ('From mother's unpublished papers', via ABC.)

Adolf HITLER 20 April 1889, 6.30 p.m. LMT, Braunau (48 N 15. 13 E 03) (From birth certificate.)

Barry HUMPHRIES 17 February 1934, 6.00 a.m. Zone 10h E, Melbourne (39 S 40. 145 E 00) (Suzanne Michaud, via *Astrological Journal,* Summer 1976.)

Carl Gustav JUNG 26 July 1875, approx. 7.30 p.m. LMT, Kesswil, Switzerland (47 N 36. 9 E 19) (7.32 p.m. 'from his daughter 'according to FCN, but FCN states Basle while *Encyclopedia Britannica* states Kesswil. ABC says Jung said he was born 'when the last rays of the setting sun lit the room'.)

Franz KAFKA 3 July 1883, 7.00 a.m. LMT, Prague (50 N 05. 14 E 22) (GAU, *Spheres of Destiny*: source not stated.)

Tom KEATING 1-2 March 1917, midnight GMT, London (50 N 30. 0 W 05) (From his autobiography *The Fake's Progress*.)

John F. KENNEDY 29 May 1917, 3.00 p.m. EST, Brookline, Mass. (42 N 21. 71 W 07) (From birth records, via FCN.)

Le CORBUSIER 6 October 1887, 9.00 p.m. Berne time (or LMT?), La Chaux de Fonds, Switzerland (47 N 06. 6 E 50) (From birth records according to FCN, though ABC says this is 'dirty data'. Neptune's position is 29°47′ Taurus, not 0°47′ Gemini as stated by FCN.)

LIBERACE 16 May 1919, 11.15 p.m. CDT, West Allis, Wisconsin

(43 N 01. 88 W 00) (From birth records, via ABC.)

John McENROE 16 February 1959, 10.30 p.m. MET, Wiesbaden (50 N 07. 8 E 17) (Richard Evans, *McEnroe, a rage for perfection*.)

Herman MELVILLE 1 August 1819, 11.30 p.m. LMT, New York (40 N 45. 73 W 57) (From family bible, via ABC.)

Eddy MERCKX 17 June 1945, 11.00 a.m. MET, Meensel-Kiezegen (assumed to be near Brussels: 50 N 51. 4 E 21) (GAU)

Henry MILLER 26 December 1891, 12.30 p.m. EST, Brooklyn, N.Y. (40 N 38. 73 W 56) (*My Friend Henry Miller*, via ABC: confirmed by GAU, *Spheres of Destiny*.)

Eric MORECAMBE 14 May 1926, 12.00 noon BST, Morecambe (54 N 05. 2 W 52) (Joan Revill in *Prediction*, November 1975.)

W.A. MOZART 27 January 1756, 8.00 p.m. LMT, Salzburg (47 N 48. 13 E 01) (From letter by native's father, via both ABC and FCN.)

Richard NIXON 9 January 1913, 9.44 p.m. PST, Yorba Linda, California (33 N 50. 117 W 46). (As recorded by midwife at time of birth, according to FCN: birth certificate says 9.30 p.m. Mars is at 29° 46′ Sagittarius, not 0° 46′ Capricorn as stated by FCN.)

Pablo PICASSO 25 October 1881, 11.15 p.m. LMT, Malaga (36 N 43. 4 W 25) (Roland Penrose, *Picasso, his Life and Work*, though ABC says this is 'dirty data'.)

Henri ROUSSEAU 21 May 1844, 1.00 a.m. LMT, Laval (48 N 04. 0 W 48) (GAU)

Satya SAI BABA 23 November 1926, exactly at sunrise (calculated for 6.09 a.m. LMT), Puttapurti (100 miles north of Bangalore: probably near Dharmavaran, 14 N 29. 77 E 44) (Howard Murphet, *Sai Baba, Man of Miracles*.)

Albert SCHWEITZER 14 January 1875, 11.50 p.m. LMT, Kayserburg, Alsace (48 N 09. 7 E 16) (Huber, from birth records, via ABC.)

Peter SELLERS 8 September 1925, 6.00 a.m. BST, Portsmouth (50 N 48. 1 W 06) (From native according to ABC; from biography according to FCN. ABC states London, but Peter Evans, *Peter Sellers, the Mask behind the Mask*, makes it clear that Sellers was born in Portsmouth. FCN correctly states 5.00 a.m. GMT, but gives positions for 5.00 a.m. BST.)

SICK CHILD 15 December 1954, Asc 27° Gemini, MC 22° Aquarius (time and place not stated). (Reinhold Ebertin, *The Contact Cosmogram*.)

Rudolf STEINER 27 February 1861, 11.15 p.m. LMT, Kraljevic (43 N 42. 20 E 43) (From his wife, via FCN: though ABC says this is 'dirty data'.)

Alan TURING 23 June 1912, 2.15 a.m. GMT, West London (51 N 32. 0 W 11) (Charles Harvey)

Lucien VAN IMPE 20 October 1946, 4.30 a.m. GMT, Mere, Belgium (assumed to be near Brussels: 50 N 51. 4 E 21) (GAU)

Charles WHITMAN 24 June 1941, Asc 2½° Taurus, MC 22¼° Capricorn (time and place not stated). (Hans-Jörg Walter, *Kosmobiologie und Sexualität,* via Ronald Harvey, *Astrological Journal,* Spring 1973. If these angles are correct, the latitude must be 27° N, and the chart has therefore been calculated for Palm Beach, Florida (26 N 46. 80 W 00) at 1.48 a.m. EST. Because of the uncertainty about this, the chart must be regarded as somewhat speculative.)

Ernie WISE 27 November 1925, 3.00 a.m. GMT, Leeds (53 N 48. 1 W 34) (Joan Revill in *Prediction,* November 1975.)

Mike YARWOOD 14 June 1941, 4.40 a.m. GMT (though BDST was in operation), Bredbury (53 N 24. 2 W 09) (Suzanne Michaud, via A.A. Data section.)

Mundane charts

Conquest of EVEREST 29 May 1953, 11.30 a.m. Zone 5½h E, 28 N 05. 86 E 58 (GBR)

Atomic bomb at HIROSHIMA 6 August 1945, 8.15 a.m. Zone 9h E, 34 N 30. 132 E 30 (J. Bronowski, *The Ascent of Man.*)

Eruption of KATMAI 1 June 1912, 1.00 p.m. Zone 10h W, 58 N 20. 154 W 59 (I cannot now trace the source for this time.)

Eruption of KRAKATOA 27 August 1883, 10.02 a.m. LMT, 6 S 10. 105 E 20 (The time of 'exactly 10.02' comes from a source which I cannot now trace, but I *think* it was W.S. Bristowe, *A Book of Islands.* GBR states '10 a.m. (local time) or 3.00 a.m. GMT': 3.00 GMT is the same as 10.02 LMT.)

Eruption of MOUNT PELÉE 8 May 1902, 7.50 a.m. LMT, 14 N 40. 61 W 00) (E.A. Vincent, in Sir Vivian Fuchs (ed.), *Forces of Nature.*)

Eruption of MOUNT ST HELENS 18 May 1980, 8.32 a.m. PDT, 46 N 29. 122 W 12 (*Sunday Times,* 25 May 1980.)

REFERENCES AND NOTES

Introduction

1 John M. Addey, *Harmonics in Astrology,* London: L.N. Fowler, 1976; also Green Bay, Wisconsin: Cambridge Circle Limited, 1976.
2 Geoffrey Dean, *Recent Advances in Natal Astrology,* Subiaco, Western Australia: Analogic, 1977, pp.137-55 and 323-32.

Chapter 1 What are Harmonics?

1 Addey, *Harmonics in Astrology,* op.cit.
2 Sheila Geddes, *The Art of Astrology,* Wellingborough: The Aquarian Press, 1980, p.156.
3 Edith Wangemann, 'The astrological aspects', *Astrology* (UK), vol. 49 no. 1, 1975. (For a discussion of Wangemann's work, see Dean, *Recent Advances in Natal Astrology,* op.cit., pp.301-2.)

Chapter 2 Which Harmonics?

1 Reinhold Ebertin, *The Combination of Stellar Influences,* Aalen: Ebertin-Verlag, 1940.

Chapter 3 Calculation and Presentation

1 James S. Williamsen, *Harmonic Chart Tables,* Green Bay, Wisconsin: Cambridge Circle Limited, 1976. (Obtainable from the Cambridge Circle at 463 Vande Hei Road, Green Bay, W1 54301, USA)
2 Addey, *Harmonics in Astrology,* op.cit., pp.108-10 of British edition; pp.101-2 of American edition.
3 Addey, op.cit., p.135 of British edition; p.130 of American edition.

4 Addey, op.cit., p.136 of British edition; p.131 of American edition.

Chapter 4 The Fourth-Harmonic Chart

1 Michel Gauquelin, *Cosmic Influences on Human Behaviour,* London: Garnstone Press, 1973, p.99.
2 Michel Gauquelin, *Spheres of Destiny,* London: J.M. Dent, 1980, p.148. (The 'famous author' referred to on this page is in fact Kafka.)
3 Anthony Storr, *The Dynamics of Creation,* Harmondsworth: Penguin, 1976, pp.78-9.
4 Gauquelin, *Spheres of Destiny,* op.cit., p.89.
5 Gauquelin, *Spheres of Destiny,* op.cit., pp.92-3.

Chapter 5 The Fifth-Harmonic Chart

1 Dean, *Recent Advances in Natal Astrology,* op.cit., pp.355-68
2 Leonard Bosman, *The Meaning and Philosophy of Numbers,* London: Rider & Co., 1974, pp.107 and 116. (Originally published in 1932.)
3 Addey, *Harmonics in Astrology,* op.cit., pp.111-2 of British edition; p.104 of American edition.
4 John Addey and Charles Harvey, 'Some suggestions on number symbolism in relation to harmonic charts', *Astrological Journal,* vol. 20 no. 1, Winter 1977-8, pp.11-13.
5 Joseph Weizenbaum, *Computer Power and Human Reason,* San Francisco: W.H. Freeman, 1976, p.58.
6 Percy A. Scholes, *The Oxford Companion to Music,* 10th edition, London: Oxford University Press, 1970, p.662.
7 Robert Hughes, *The Shock of the New: Art and the Century of Change,* London: BBC, 1980, p.146.
8 Jonathan Raban, 'The old man and the word', *The Sunday Times,* 26 April 1981. (Review of Carlos Baker (ed.), *Ernest Hemingway: Selected Letters, 1917-1961,* Granada.)
9 Storr, *The Dynamics of Creation,* op.cit., p.122.
10 Jean Cocteau, in: John Wilson (ed.), *The Faith of an Artist,* London: Allen & Unwin, 1962, p.83. (Quoted in Storr, op.cit., p.124.)
11 Reinhold Ebertin, *The Contact Cosmogram,* Aalen: Ebertin-Verlag, 1974, pp.105-6 (chart on p.108).

Chapter 6 The Seventh-Harmonic Chart

1 Charles Harvey, 'Some suggested interpretations for planets when prominent in the seventh-harmonic chart', unpublished manuscript, 1980.
2 Charles M. Graham, 'The seventh harmonic and creative artists', *Astrology Now,* no.13, June 1976. (Also published as a typescript by Cambridge Circle Limited, together with a second article 'The creative impulse: practical use of the seventh harmonic'.)
3 For instance: Eric Newton, *The Meaning of Beauty,* London: Longmans Green, 1959; Robert Pirsig, *Zen and the Art of Motorcyle Maintenance,* London: Bodley Head, 1974.
4 Scholes, *The Oxford Companion to Music,* op.cit., p.106.
5 *Phaidon Dictionary of Twentieth-Century Art,* 2nd edition, London: Phaidon Press, 1977, pp.108-9.
6 Hughes, *The Shock of the New,* op.cit., p.241.
7 *Phaidon Dictionary,* op.cit., p.86.
8 Rupert Brooke, 'Vanitas', in: Geoffrey Keynes (ed.), *The Poetical Works of Rupert Brooke,* London: Faber & Faber, 1946, pp.170-1.
9 I cannot now identify the precise source for this statement, but I believe that it was obtained from a *Sunday Times* review of a recent biography of Henry Miller.
10 Le Corbusier, quoted in Hughes, op.cit., p.188.
11 Charles Harvey, 'Harmonic charts', *Astrological Journal* vol.20 no.1, Winter 1977-8, pp.2-11.

Chapter 7 The Ninth-Harmonic Chart

1 Gauquelin, *Spheres of Destiny,* op.cit., p.137.
2 Suzanne Lilley-Harvey, 'Rudolph Steiner', *Astrological Journal,* vol.22 no.2, Spring 1980, pp.72-7.
3 Rudolf Steiner, *Knowledge of the Higher Worlds: How is it Achieved?* , translated by D.S. Osmond and C. Davy, revised edition, London: Rudolf Steiner Press, 1969.
4 Alan Oken, *The Horoscope: the Road and its Travelers,* New York: Bantam Books, 1974, p.376.
5 Addey, *Harmonics in Astrology,* op.cit., p.103 of British edition; pp.95-6 of American edition.
6 Addey, loc.cit.
7 Bernard Dorival, 'The Realist Movement', in: René Huyghe (ed.), *Larousse Encyclopedia of Modern Art,* London:

Hamlyn, 1965, pp.159-74. Quotation from p.167.
8 Gäeton Picon, *Modern Painting from 1800 to the Present,* New York: Newsweek Books, 1974, p.46.
9 Charles Eliot Norton, in *North American Review,* April 1868. (Quoted in: J. Wintle and R. Kenin (eds.), *The Dictionary of Biographical Quotation of British and American Subjects,* London: Routledge & Kegan Paul, 1978, p.237.)
10 Charles Dickens, in a letter to J.V. Staples, 3 April 1844. (Quoted in Wintle and Kenin, op.cit., p.235.)
11 Peter Evans, *Peter Sellers, the Mask behind the Mask,* revised edition, London: New English Library, 1980, p.194.

Chapter 8 General Hints on Interpretation

1 Dean, *Recent Advances in Natal Astrology,* op.cit., pp. 355-68.
2 Addey, *Harmonics in Astrology,* op.cit., p.127 of British edition; p.120 of American edition.

Chapter 9 The Planets

1 David Hamblin, 'Hot and cold murder', *Astrological Journal,* vol.16 no.1, Winter 1973-4, pp.29-36.

Chapter 12 Eleonora Duse

1 Gauquelin, *Cosmic Influences on Human Behaviour,* op.cit., pp.110-11.
2 Dean, *Recent Advances in Natal Astrology,* op.cit., p.364.

Chapter 13 Henri Rousseau

1 Ronald Alley, *Portrait of a Primitive: the Art of Henri Rousseau,* Oxford: Phaidon Press, 1978, pp.67-8.
2 Alley, op.cit., pp.70-1.
3 Alley, op.cit., pp.49-50.
4 Alley, op.cit., p.62.
5 Modest Morariu, *Douanier Rousseau,* London: Abbey Library, 1979, p.20.

Chapter 14 Ludwig van Beethoven

1 *Transit* (the Astrological Association Newsletter), issue no 36, January 1982, p.14.'

2 Scholes, *The Oxford Companion to Music,* op.cit., p.93.
3 Scholes, op.cit., Plate 14 (facing p.92).
4 J.W.N. Sullivan, *Beethoven,* Harmondsworth: Penguin, 1949, p.48.
5 Sullivan, op.cit., p.46.
6 Sullivan, op.cit., p.147.

Chapter 16 Susan

1 Ebertin, *The Combination of Stellar Influences,* op.cit., p.101.

Chapter 17 Harmonic Synastry

1 Lois A. Rodden, *Profiles of women: a collection of astrological biographies,* Tempe, Arizona: American Federation of Astrologers, 1979, p.132.
2 Robert Hand, *Planets in composite: analyzing human relationships,* Gloucester, Mass.: Para Research, 1975.
3 Ebertin, *The Combination of Stellar Influences,* op.cit.
4 Ronald Davison, *Synastry: understanding human relations through astrology,* New York: ASI Publishers, 1977.

Chapter 18 Harmonic Charts for Mundane Events

1 Information from the *Guinness Book of Records*; also from E.A. Vincent, 'Volcanoes', in: Sir Vivian Fuchs (ed.) *Forces of Nature,* New York: Holt, Rinehart & Winston, 1977, pp.210-32.

Chapter 19 Harmonic Transits

1 Roland Penrose, *Picasso: his life and work,* 3rd edition, London: Granada, 1981, p.301.
2 Since this passage was written, it has been rumoured that Brady may be about to confess to two other murders committed before this date; but at the time of writing he has not done so.
3 Dianne Binnington, letter in *Astrological Journal,* vol.24 no.1, Winter 1981-2, pp.54-5.
4 Addey, *Harmonics in Astrology,* op.cit., pp.165-7 of British edition; pp.161-3 of American edition.

Chapter 20 Other Harmonic Numbers

1 Harvey, 'Harmonic charts', op.cit.
2 Addey and Harvey, 'Some suggestions on number symbolism in relation to harmonic charts', op.cit.
3 John Addey, 'Seven thousand doctors', in *Selected Writings,* Tempe, Arizona: American Federation of Astrologers, 1976, pp.189-201.
4 John Addey, 'Fivefold division and sub-divisions in astrology', in *Selected Writings*, op.cit., pp.214-28.
5 Martin Seymour-Smith, *The New Astrologer*, London: Sidgwick & Jackson, 1981, pp.125-6.
6 Greta Garbo, quoted in Oken, *The Horoscope*, op.cit., p.364.
7 Seymour-Smith. op.cit., p.125.
8 Quotations from Evans, *Peter Sellers, the Mask behind the Mask,* op.cit., p.171.

Chapter 21 Harmonics of the Zodiacal Circle

1 Addey, 'Seven thousand doctors', op.cit., also *Harmonics in Astrology,* op.cit., pp.202-3 of British edition; pp.199-201 of American edition.
2 Figure 10 is derived from the table of the position of the Sun by degrees for the 7302 doctors of medicine, given by Addey in 'Seven thousand doctors', op.cit., p.199.
3 Addey, 'Seven thousand doctors', op.cit.
4 Reported in Addey, *Harmonics in Astrology,* op.cit., pp.124-5 of British edition; pp.117-8 of American edition.
5 Addey, 'Seven thousand doctors', op.cit., p.191.
6 Bruno and Louise Huber, *Man and His World: astrological psychology,* New York: Samuel Weiser, 1978. (This book is more concerned with houses than with signs, and does not contain a clear statement of a theory of overlapping signs, but I believe that it is based on such a theory. The Hubers believe that the *intensity* of each sign is greatest at around 12° of the sign.)
7 John E. Greig, *Astrology and Planetary Harmonics,* London: Astrological Association, 1980.

Chapter 22 Harmonics of the Diurnal Circle

1 This statement is not strictly correct, since the Ascendant and Descendant are opposite to each other in space but not

in time. In the diurnal circle as measured by Gauquelin, both the time taken by a planet to cross from Ascendant to Descendant (above the horizon) and the time taken to cross from Descendant to Ascendant (below the horizon) are each counted as 180°, even though one may in fact (except at the equator) be greater than the other. This is clearly shown in the diagram in Dean, *Recent Advances in Natal Astrology,* op.cit., p.382.

2 Astro Computing Services (director: Neil F. Michelsen), P.O. Box 16297, San Diego, CA 92116, USA.

3 Gauquelin, *Cosmic Influences on Human Behaviour,* op.cit.; *Spheres of Destiny,* op.cit.; and other works. Details of the Gauquelins' scientific papers and lists of birth data can be obtained from them at 8 Rue Amyot, 75005 Paris, France.

4 See especially Dean, *Recent Advances in Natal Astrology,* op.cit., pp.380-94.

5 Gauquelin, *Spheres of Destiny,* op.cit.

6 See Dean, op.cit., pp.1-5; also H.J. Eysenck and D.K.B. Nias, *Astrology: Science or Superstition?,* London: Maurice Temple Smith, 1982; also my own article 'The need for doubt and the need for wonder' (see Chapter 23, ref.1).

7 Frances Sakoian and Louis S. Acker, *The Astrologer's Handbook,* London: Peter Davies, 1974, pp.154 and 161.

8 Michel Gauquelin, *Scientific Documents no.9: Murderers and Psychotics, with the publication of 7000 birthdata,* Paris: Laboratoire d'étude des relations entre rythmes cosmiques et psychophysiologues, 1981.

9 John Addey, 'Harmonic phase and personal characteristics: Part I', *Astrological Journal,* vol.21 no.3, Summer 1979, pp.123-34; 'Harmonic phase and personal characteristics: Part II', *Astrological Journal,* vol.22 no.1, Winter 1979-80, pp.7-13 and 52.

10 Addey, 'Harmonic phase and personal characteristics: Part I', op.cit., pp.133-4.

11 John Addey, 'The basic harmonics of Mars and their interpretative value in the horoscope: Part I, The first and second harmonics', *Astrological Journal,* vol.23 no.4, Autumn 1981, pp.208-16; 'Part II, The third and fourth harmonics', *Astrological Journal,* vol.24 no.1, Winter 1981-2, pp.12-20.

12 Addey, 'Part II, The third and fourth harmonics', op.cit., p.15.

13 Addey, 'Harmonic phase and personal characteristics, Part II', op.cit.

Chapter 23 The Search for Truth

1 David Hamblin, 'The need for doubt and the need for wonder', *Astrological Journal,* vol.24 no.3, Summer 1982, pp.152-7.
2 Dean, *Recent Advances in Natal Astrology,* op.cit.
3 Michael Polanyi, *Personal Knowledge: Towards a Post-Critical Philosophy,* London: Routledge & Kegan Paul, 1958, p.64.
4 Polanyi, op.cit., p.135.
5 Zipporah Pottenger Dobyns, *Finding the Person in the Horoscope,* Los Angeles: T.I.A. Publications, 1973.
6 Margaret F. Hone, *The Modern Textbook of Astrology,* London: L.N. Fowler, 1951, p.181.
7 Addey, *Harmonics in Astrology,* op.cit., p.18 of British edition; p.11 of American edition.
8 Ebertin, *The Combination of Stellar Influences,* op.cit.
9 *Ecclesiastes,* 1.vii.
10 For instance: John Addey, 'The discovery of the scientific starting-point in astrology', in *Selected Writings,* op.cit., pp.173-88.
11 Dean, *Recent Advances in Natal Astrology,* op.cit., pp.433-4; also pp.190-2 for a description of Blunsdon's 3D chart which takes latitude into account.
12 Dean, op.cit., pp.281-2; but for calculating parallax the table from the following publication should be used (as the table in *Recent Advances* is incorrect): Geoffrey Dean, 'Corrections for lunar parallax', *F.A.A. Journal,* vol.4 no.2, April 1974, pp.5-9.
13 Personal correspondence to the author from Geoffrey Dean and from James MacPherson.
14 Hamblin, 'The need for doubt and the need for wonder', op.cit.
15 Polanyi, *Personal Knowledge,* op.cit., pp.6-9.
16 A duodecimal system is one in which counting is done in twelves rather than tens. Two extra digits (which we can call plink and plonk) are created to represent ten and eleven. Twelve becomes ten, twenty-four becomes twenty, 144 becomes a hundred, and 143 becomes plonkty-plonk.
17 Olaf Stapledon, *Last and First Men,* Harmondsworth: Penguin, 1937, pp.287-8.

INDEX

SPECIAL NOTE: It is planned to start an international Harmonic Astrology Newsletter, for the exchange of information about the latest developments in Harmonics, and also to allow astrologers to share their experiences of using Harmonics in their practical work. If you would like to know more about this, please write to David Hamblin, c/o 36 Tweedy Road, Bromley, Kent, BR1, 3PP, U.K.